CARGO
3120

CARGO 3120

TIES THAT BIND

AARON WALKER SR.

ARPress

ILLUMINATING IDEAS.
EMPOWERING VOICES

ARPress
45 Dan Road Suite 5
Canton MA 02021

Hotline: 1(800) 220-7660
Fax: 1(855) 752-6001

Ordering Information:
Quantity sales. Special discounts are available on quantity purchases by corporations, associations, and others. For details, contact the publisher at the address above.

Printed in the United States of America.

ISBN-13:	Paperback	979-8-89330-710-8
	eBook	979-8-89330-711-5
	Hardcover	979-8-89330-712-2

Library of Congress Control Number: 2024901777

To my Wife, Terrie. For your unfailing support and for
all the nights you stayed up, listening to my early drafts.
I love you and thank you for believing in me.

To my children, No'Elle, Aaron Jr., Elias, and to the rest of
my family, I love you all very much. Thank you for being
who you are, and for your support from the beginning.

To the co-creators of Cargo 3120, Daymond C.
Roman and Lloyd Walker Jr. Without you guys,
this universe would have never come to life.

To Larry Brody. Thank you for being an early believer in this
project. I can't thank you enough for sharing with me your wealth
of storytelling knowledge and experience over the years.

Finally, I thank God for giving me the ability to write.
Without him, none of this would be possible...

*"Every good and perfect gift is from above, coming down from the Father
of the heavenly lights, who does not change like shifting shadows."*

—James 1:17

TABLE OF CONTENTS

PART 1: THE TITAN JOB

PART 2: RUNNER'S END

PART 3: A CLASH OF DESTINIES

PART 4: THE DARK TEMPEST

PART ONE

THE TITAN JOB

"I hope it was worth it."

— Skyela "Skye" Evans

1

CHAPTER ONE

As Marcus La'Dek and his partner, Daren West, negotiated the muddy road leading to the target mining facility, Marcus remembered just how much he hated the rain. The ground had been reduced to black sludge, thanks to the torrential downpour battering the moon's northern hemisphere for the past twelve hours. But the deluge brought with it more than mere sodden conditions. A ghostly mist of strange particulates always accompanied the rains, making it difficult to see over five feet in any direction. Therefore, the use of advanced image processing goggles, known as IPGs, was a necessity.

Except for an annoying yellow hue, Marcus felt the goggles provided remarkable clarity. But if given the choice, he would rather have donned his own tactical shades, as the built-in targeting optics and image-enhancement technology outperformed the IPGs in every way. But alas, blending in was the name of the game for this mission, so he was okay with settling for less.

"Particle storms" are what the inhabitants of the moon called the atmospheric phenomenon. Many experts attributed the mist to the failing methane scrubbers of Titan's problematic Terraforming Modules. Although further research was needed to be sure, it was unlikely that the human-led Interstellar League of Planets (ISL) would spend the time or money on such a study. For Titan, Saturn's largest ISL-controlled moon, was dedicated to one purpose, mining the lifeblood of terraformed planets and moons across the galaxy, Krillium ore. And as long as environmental conditions on the surface didn't cause the moon's populace to drop dead in droves, business as usual would continue, regardless of the many complaints lodged with the ISL government.

Taxpayer dollars hard at work, Marcus thought as he adjusted the hood of the black poncho he wore over the tactical gear, which clothed his brown, muscular frame. Given the circumstances, Marcus and his team couldn't ask for better conditions for a nighttime raid on a guarded mining facility to plunder a drove of that precious Krillium. Sure, it was a dangerous business, but the substantial sum of money for which their employer would pay made the risk worth the reward.

Marcus and his small crew of mercenaries considered themselves to be little more than "mules," occasionally picking up and dropping off ore from hard-to-infiltrate facilities on behalf of their employer, Teric Winters, the reclusive leader of the Outer Core's largest and deadliest terrorist organization, Orion's Shield.

What could he say? The pay was good, and the job allowed him to travel to exotic locations.

But not today, Marcus thought as he and Daren, his number two, approached Titan's largest ore-processing facility, ISL Mining Outpost Alpha.

"Here we go. Just follow my lead," said Marcus as they neared the military checkpoint that guarded the entrance to the complex.

The checkpoint consisted of a small control booth attached to a larger guard tower. But that's all the facility needed in the way of protection. The entire complex was surrounded by a translucent green energy shield that would take nothing less than a starship bombardment to breach.

"Why don't you let me handle this one?" Daren asked. "Situations like these require a little more, finesse. Ya know?"

"Be my guest," Marcus answered. "But don't screw this up. We only get one shot at this."

As they neared the checkpoint, the storm seemed to kick into high gear, causing their ponchos to flap like flags in the wind. But despite the harsh conditions, they pressed forward, confident their plan would go off without a hitch. They had it all worked out. Prior to reaching the main road, three members of Raven Squad—Heavy weapon's specialist Tony Liles, computer specialist and Combat Medic Skyela "Skye" Evans, and Explosives expert Maxlyn "Max" Wesner—fell out of formation, taking up positions behind several large boulders, awaiting their cue to assist in the heist.

#

While Tony, Skye, and Max awaited their orders from Marcus and Daren, the sixth and final member of Raven Squad, Jason Crowley, was posted several miles back at their ship, *The Indicator*, which he had landed on a small rocky plateau.

Named after the long extinct Earth fowl, *The Indicator* was an experimental, medium-sized gunship modified from a cargo vessel. It housed an impressive collection of advanced sensors and astrometric locator equipment and was armed to the teeth.

Sitting at *The Indicator*'s helm with his feet on the navigation console, Jason had been using the ship's electromagnetic pulse emitter to send a discrete signal toward Mining Outpost Alpha for the past several hours. The goal was to surreptitiously scramble local communication and security sensors.

"The art of jamming without jamming" is what Jason called the complex process. He knew his low-level pulse would go virtually undetected, and the interference would likely be attributed to both the storm and aging security equipment. After checking the integrity of the signal one last time, Jason leaned back in his chair, confident that nothing could possibly go wrong.

#

Near the mining outpost, Marcus and Daren emerged from the mist like shadowy apparitions, with only the faint yellowish glow of their IPGs visible to the guards manning the checkpoint. Yet as they approached, the guards had no cause for alarm, believing Marcus and Daren to be fellow soldiers from the nearby work camp.

"Sorry, guys. We're gonna have to do a manual security check," one guard said. "The automated system is down, must be this freak particle storm."

"Of course it's down," said Daren. "Why do you think we're here?" As the guards looked to each other confused, Daren continued. "Don't worry about it. The watch commander sent us to fix the comm and security systems. So, if you'll kindly let us through..."

"What are you talkin' about? They already have guys inside workin' on the system," said another guard.

"Well, if those 'guys' knew what they were doin', we wouldn't be here, now would we?" said Daren. He went back and forth with the guard for several minutes trying to manipulate his way inside the facility.

In reality, talking was Daren's forte. Standing six-foot-two, with a slender yet muscular build and dark brown skin, Daren often got by on his rugged good looks and silver tongue. He often used his verbal prowess to get in and out of situations, not to mention, to score with the ladies.

But to Marcus, Daren seemed off his game. Either that, or these guards were a lot brighter than Daren anticipated.

"C'mon, guys, give us a pass," Daren continued. "Today was our day off. How do you think we feel?"

#

From the guard tower, a fourth soldier peered through the thick ballistic glass window, eying the situation on the ground. He couldn't hear the conversation because of the ridiculous level of static blaring through his speakers. *I hate this place. Nothin' ever works around here*, the guard thought. He switched off the useless speakers and addressed his growling stomach instead, figuring everything to be under control on the ground. So, he sat down and stuffed his face with the various treats he had smuggled into the tower before the start of his shift.

#

"The Commander's already pissed at us for last week. So, no passes today," said the guard with whom Daren had been speaking.

Annoyed with Daren's ruse, which at present was spiraling to a flaming wreck, Marcus used the distraction to assess the threat level. Whereas Daren preferred words, combat was where Marcus was most comfortable. Through years of rigorous training in several styles of martial arts, Marcus considered himself an aficionado in the ways of doling out pain. But his skills didn't stop at melee. He could also shoot the wings off a Kor'Dalean fruit fly with a pistol from two hundred meters.

Marcus could tell these guys didn't see much action either. Guard number one, the soldier arguing with Daren, was armed only with an

AP17 tactical pistol, standard issue for ISG soldiers. His rifle was likely sitting next to the box of Lyrian puff pastries Marcus observed inside the control booth.

And instead of manning their posts, guards two and three stood behind the first with their AX19 assault rifles casually slung over their shoulders, instead of at the ready. They stood snickering at the verbal exchange between Daren and their squad leader. He thought at least the tower guard would be more observant. But Marcus wrote him off as a non-threat after he noticed the soldier taking his ill-timed lunch break. Too easy, Marcus thought.

Marcus slowly reached beneath his poncho toward the two XP-90s holstered to his belt, his favorite pistol. Its design was sleek, and appearance sexy, but they were more than just pieces of ostentation. Each pistol carried a multi-layered magazine with various specialty rounds. Using controls on the pistol's grip, Marcus could change each weapon's fire mode on the fly. Its most versatile load comprised ten standard ballistic rounds and five shock rounds, for a less lethal approach. And when changed to the airburst setting, a nerve round could punch through a wall and detonate midair, dispersing its gas in seconds.

Daren noticed the hard look Marcus was giving him.

"I'm tellin' you," Daren yelled to guard number one, "we have orders from the watch commander himself..."

As Daren continued with the guard, a green hover jeep made its way up the muddy road toward the checkpoint. Another ISG soldier exited the vehicle with a cup holder containing four drinks of what Marcus figured to be coffee. The soldier was a scrawny man wearing a uniform that looked to be two sizes too big. He had no facial hair and wore thick black glasses. Even though the 32nd century offered many options to correct eyesight deficiencies, many humans favored the retro look of the 21st century. So, they adorned these superfluous spectacles merely for the fashion benefits they offered.

"What's goin' on here?" the new arrival asked. "Who are you guys?"

Daren ignored the java-toting soldier and continued with the guard. "Look, genius, you think I'm out here in the pouring rain for my health? The commander ordered us to—"

"Hey, I'm talkin' to you, soldier," the skinny visitor interrupted.

Daren finally turned toward the source of his annoyance, and after one look, counted the guy to be some fresh-out-of-boot-camp errand boy.

"Look, junior, piss off," Daren shot back. He looked down at the beverages and changed his mind. "Better yet, run and grab two more of those, will ya?"

Furious at Daren's audacity, the soldier threw the cups to the ground. "Just who do you think you're talkin' to? I'm the Commander of the Watch."

A knot swelled in Daren's throat, though he did a remarkable job of concealing that fact. He and Marcus knew that Daren's ill-conceived plan had just derailed.

"Oh, Commander... We... I mean, I was just sayin'," Daren said, attempting to recover.

Having seen and heard enough, Marcus acted before the freak show grew any worse. He tightened his grip on those two XP-90s beneath his poncho. He set the left pistol to its shock round setting and the other to nerve gas. Then, Marcus pulled the two weapons out, simultaneously firing at the watch commander on the ground and the lone guard in the tower.

A cylindrical round blasted the watch commander's chest, delivering a high voltage, low current shock that sent the man to the ground in a violent convulsion. The second round crashed through the ballistic window of the guard tower. Before the half-asleep guard could react to the sound, the nerve round detonated midair, filling the room with its toxic vapor. The gas worked quickly, sending the guard crashing to the floor into what looked like a violent seizure.

With one swift motion, Marcus holstered the sidearm in his left hand, and flipped the other into the air. He grabbed the pistol's barrel in midair and used its grip to strike guard number one in the face, knocking the unprepared man unconscious. Next, Marcus turned his attention to the last two guards who nervously struggled to unsling their rifles.

Marcus holstered his second sidearm and engaged the over-matched soldiers in close-quarters combat, delivering a brutal series of punches, knee strikes and elbows. The assault lasted but a few seconds, leaving both guards barely conscious and with broken limbs. The entire ordeal happened so fast that even Daren was unable to join the fray.

"Really, Daren?" Marcus said while pointing to the watch commander on the ground. "You didn't see that guy's rank?"

"I had it under control."

"We don't have all night. Now clean this up."

Realizing the futility of arguing with Marcus, Daren compiled the incapacitated guards into the control booth, after which he bound and gagged them. Under normal circumstances, they would have just shot the guards and dumped the bodies. But when raiding an ISL controlled facility, the last thing you needed were murder charges added to your list of crimes if caught. Therefore, Marcus put a zero-body-count rule of engagement into effect, with no exceptions.

As Daren secured the guards, Marcus contacted Jason on *The Indicator*. He pressed a button on his earpiece to activate his comm unit. "Raven six, this is Raven Leader. We're in position."

Jason responded over *The Indicator*'s radio, "Leader, this is Six. The jamming signal is holding. I'm sure they're havin' a hell of a time findin' the problem."

"Copy that, Six. Standby for the extraction order."

Marcus approached the control panel that operated the energy shield. Months of research and planning had paid off. He bypassed the shield's security measures with a small data stick containing a custom-written virus, courtesy of Raven Squad's resident computer genius, Skye.

"We're in business," Marcus said to Daren. "You ready?"

"Almost. Just need to secure our buddy in the tower. Go ahead. I'll be there in a second."

"Whatever. Just make it fast," Marcus said to his partner as he exited the control booth toward the unknown opposition awaiting them inside the mining facility.

CHAPTER TWO

Once Marcus left the control booth, Daren proceeded to the tower to secure the last guard. Upon reaching the lofty structure, he pulled up the facemask portion of his balaclava to cover his nose and mouth and took the express lift to the top floor. The mask was made of a special fabric designed to filter harmful substances lingering in the air. He thought it best not to take chances, lest he himself should end up like the poor guy lying inside.

After binding and gagging the already unconscious guard, Daren used his personal data pad to contact an associate with whom he was working a small side job.

While the team's official mission was to steal crates of raw Krillium ore, Daren and his associate were after its refined counterpart. After a few holographic button presses, Daren's data pad came to life with the scrambled image of Tommy Lance, a battle-hardened mercenary operating under the call sign "Specter One."

Tommy was the most gifted sniper in the entire Orion's Shield organization. And when it came to underworld contacts, he had no equal. While preferring to work alone, Tommy desperately needed help to boost the refined ore for one of his important contacts. So, when he heard Raven Squad was to hit the mining outposts on Titan, Tommy called in a favor with Daren, his longtime business associate.

While stealing the rare mineral in its raw form, Daren expected little to no resistance, as the ore storage area was free of any guard presence. Of course, boosting the refined Krillium stockpiles posed the biggest problem as its security was much tighter. Ever the opportunist, Daren was all too eager to accept Tommy's request, seeing the opportunity as a chance of a lifetime, a chance to finally step out of the long shadow cast by the legendary Marcus La'Dek.

Tommy promised him wealth beyond comprehension and entry into a shadow organization known only as The Company. But Daren wanted more than just entry into an organization. He was tired of being "number two." But it was not a decision lightly made. For Daren's loyalty to Marcus was strong, having lived together since they were children.

Even Daren's half-sister and original team leader, Dayna West, accepted Marcus as a brother during the time he lived and trained alongside the siblings by their father, the renowned Outer Core pirate, Johnathan "Jax" West. Even after Daren's sister abandoned the team years later, a painful subject they seldom discussed, Marcus and Daren picked up the shattered remains and pressed forward, inseparable as if they were blood brothers.

"Tell me you have somethin'," Daren said while peering into the data pad at his silent partner.

"Relax, man, you're in luck. The refined Krillium is near your team's target crates. But it's guarded by a RAT," Tommy replied in a calm voice.

Daren couldn't believe what he was hearing, feeling the mission was already going downhill before it even began. When Tommy mentioned the word RAT, Daren knew he wasn't referring to the common form of rodentia found on Earth. Weighing in at seven tons, the Robotic Autonomous Tank, or R.A.T., was a heavily armed weapon of extreme death and destruction, featuring multiple nose cannons and two energy-based Gatling guns on the upper left and right sides of its chassis. But the defining feature of the RAT was that it was bipedal, with some designs capable of walking on all fours, making them agile, capable of operating on just about any terrain.

"A RAT? Oh, hell no, you do it," Daren replied. Having second thoughts, Daren knew there was more to fear than the metal behemoth. "Plus, if Marcus finds out? That walkin' tank will be the least of our—"

"Would you man up already?" Tommy interrupted. "This is our only shot to get in with The Company."

Daren shuddered at the mention of The Company. The organization's name invoked both exhilaration and fear, so he hesitated. Part of him wanted to call the whole thing off. How could he jeopardize the mission and the entire team without Marcus' knowledge or approval? How could he betray the only person in his life whom he considered being family? But Daren's desire to forge his own path and to be a person of great import was too strong. He was, as was so often the case, his own worst enemy. So, he conceded with a nod of agreement.

"My man. Time to step out of the shadows, brotha. Specter One, out," Tommy said. Satisfied that he had convinced his partner in crime to stay the course, Tommy ended the communication.

Daren stared into the blank screen of his data pad, seeing only his faint reflection. And he didn't like what stared back.

#

Hidden several miles outside the front gate behind large boulders, Max, Skye, and Tony sat awaiting the go order. While Max and Skye made small talk amongst themselves, Tony, as he often did, showed little interest in the women's chatter.

"Say what you want, this don't feel right," Max said to Skye while fidgeting with a small pebble she found nearby.

Max was always nervous about doing jobs near Earth, her former home world. Hailing from the North American continent in the state of Texas, Maxlyn Wesner was a simple country girl with a southern drawl and a love for explosives. A brilliant scientist holding dual PhD's in chemistry and math, Max started her career designing explosive ordnances for The Maxis Corporation, a multi-trillion-dollar human-run defense contractor.

However, Max's promising career ended after she caught her boyfriend sleeping with her own cousin. Max blew up his Las Vegas penthouse and fled Earth on a one-way ticket to the Outer Core, often called The OC, never again to return. So, even operating in the same star system as the planet Earth brought back a flood of anxiety.

"C'mon, Max. Not this again," Skye replied. She looked over at Tony, who seemed bored out of his mind. He kept peeking from the corner of the boulder toward the mining facility in the distance. The complex was partially obscured from their view because of the heavy mist. All they could see were faint lights emanating from the buildings of the compound.

"Hey, Tony," Skye called out. "Isn't it your turn to deal with Maxlyn's butterflies?"

"I'm tired of sittin' here playin' in the dirt. The shield's down. What's taking them so long?" Tony responded without addressing Skye's question.

"Seriously, guys," said Max, "We've never hit a facility this close to Earth before."

"Look, there's nothin' to worry about," Skye assured her comrade. "No one's gonna recognize you out here. Plus, the security at this place is... less than vigilant."

"And tell me, darlin', just how would you know that?" Max asked in her heavy Texas accent.

"The ISG stationed me here just before they gave me the boot," Skye admitted while checking her pistol to ensure its readiness. "Most of the soldiers on this backwater moon are a bunch of rejects and losers on their way out of The Guard."

Skyela Evans had smooth, dark skin, full lips, and slightly curly, shoulder-length black hair, which she kept tightly braided during missions. Tough as nails, and a prodigy with computers and electronics, Skye received extensive training as a combat medic during her brief stint with the Interstellar Guard (ISG), the military ground force of the Interstellar League.

Skye explained to Max how she was forced to work gate security for none other than their current target—Mining Outpost Alpha—while awaiting her dishonorable discharge from ISG service. So, Skye assured Max that her extensive knowledge of the facility gave them a distinct advantage.

"Is that right?" Max replied after hearing Skye's story. "Still don't feel right."

"Just focus on your job and try not to blow us up out there," Skye said as she holstered the sidearm.

Max smiled as she carried the dubious distinction of going overboard with her elegantly crafted explosives. Yet there was no denying the genius of those designs. But before she could respond, the call they had all been waiting for came through.

"Wake up, Ravens, time to move," Daren's voice crackled over the comms.

Tony was first to spring to his feet. "Finally," he said, not even attempting to hide the excitement in his voice. He looked over at Skye and Max. They didn't move with the same sense of urgency. "You heard the man, let's move."

Tony Liles was always eager for a fight. It was in his nature, literally. Despite his human name, Tony was a Gorean Cyborg, raised by humans on the war-torn Outer Core planet, Bion IV. Goreans were a genetically engineered humanoid race, bred for combat.

Born part man, part machine, within breeding chambers, Goreans had eyes like faintly glowing orbs with no pupil. The Goreans' mostly subdermal cybernetics could only be described as a living metal, as much a part of their anatomy as a human's skin and bones.

Though anatomically similar to humans, Goreans were much taller, with broad shoulders, enlarged trapezius muscles, and a partially exposed metallic spinal column serving as part of their natural exoskeletal armor.

With massive arms of flesh and living metal, Tony grabbed his gear and prized heavy repeater, a large energy-based machine gun designed for suppression and destruction. It was time to go to work.

Skye and Max grabbed their packs and stood to their feet. But Max still looked uneasy as she secured her long, sandy-blonde hair beneath her balaclava. As they converged on Outpost Alpha, Max trailed behind, hoping for the best, but expecting the worst.

CHAPTER THREE

Inside the ore-processing center, Marcus was faring just fine despite his partner's extreme tardiness. If there was one thing that irritated him about Daren, it was his lack of a strong work ethic. Daren was all talk with his grandiose ideas and dreams, but never putting forth the effort required to make those things happen. Marcus felt that Daren spent too much time feeling sorry for himself, while drinking and getting high on zeth—a popular narcotic from the OC. But Daren was a skilled fighter and fiercely loyal, so he often looked past his partner's faults. But at that moment, he wished Daren would hurry.

With silent, calculated strides, Marcus approached what turned out to be the administrative wing of the building, filled with clusters of computer stations manned by underpaid office personnel forced to work the late shift. Armed guards patrolled the multi-leveled area.

Marcus felt the prudent action would be to talk his way into the security room, but that was more Daren's thing. Besides, even if he made it inside, he wouldn't be able to do anything since Skye wasn't present to disable the system. Daren had only two jobs: secure the guards, call in the team. And he couldn't even get that right. Once again, Marcus would have to take matters into his own hands. So, he studied the room, paying close attention to the movement patterns of the guards, when he hit a snag.

"Hey, you," called one worker who seemed to come out of nowhere. "The latrine is out of commission again. Come over here and give us a hand."

For a moment, Marcus wished Daren were there to work his verbal virtuosity. But the mission was already behind schedule, so he would have to neutralize the entire room himself. Since internal communications and alarms were still down, thanks to Jason's electronic masterwork, Marcus figured he could pull this off on his own. The trick was making sure no one left the room to alert the others.

"Hey, man. Somethin' wrong with your ears?" the worker asked as he approached from behind. He tapped Marcus' shoulder.

"Yeah, sure, be right there," Marcus replied. Without turning around, Marcus caught the unsuspecting worker with a swift elbow to the chin, knocking him cold. With quick reflexes, he turned and caught the victim before his body hit the ground.

After scanning the room to make sure his deeds remained unseen, Marcus dragged the guard behind a nearby desk, out of view of the rest of the room.

Marcus ditched the poncho, exposing his black tactical gear, filled with pouches of ammo clips, grenades of various types, and a serrated combat knife. He swapped out the IPGs for his own tactical goggles from his pack.

Still crouched behind the desk, Marcus peeked around the corner, scanning the room with his electronic eyewear. From his perspective, the glasses identified and tagged all hostile targets, while also providing detailed data on every weapon carried by the soldiers.

As far as the noncombatants, Marcus was in luck as they were all grouped closely, working at two separate clusters of workstations, easily within the blast radius of his grenades.

He estimated two would do the trick. From his equipment pouch, he grabbed twin grenades filled with the same nerve agent that had dropped the tower guard. Marcus set them to explode on impact and pulled the pins.

Covering his nose and mouth with his balaclava, Marcus propelled the grenades toward the civilians with consecutive underhand tosses. With any luck, he hoped to catch a stray guard or two. He was right. Just as the grenades reached their targets, one guard walked into the blast zone. With loud bangs, the grenades went off, startling everyone in the room. Before the people on the floor could move, the thick vapor enveloped them, causing everyone to fall to the floor, immobilized.

"Gas," yelled one of the guards on the upper level to the rest of the room. Those unaffected by the fumes donned their gas masks and IPGs, searching the area for the threat.

No sooner than they could turn their heads, many of them dropped like flies as Marcus nailed them with shock rounds from his rapid fire XR75 assault rifle. Several guards emerged from a nearby break room trying to blast the intruder with their weapons. But Marcus proved to be an elusive target, diving in and out of cover, switching between his rifle and pistol, taking down all in his path.

Marcus lived for those moments. Though outnumbered, the guards and remaining civilians never stood a chance. The few soldiers unfortunate enough to get close, he engaged in hand-to-hand combat. Marcus was a master of using his environment to his advantage. Using bare hands, tables, pipes, and anything else he could get his hands on, Marcus pressed the attack with almost unnatural speed. Those looking to make an unceremonious exit found themselves shot in the back with a fresh salvo of shock rounds before they could reach the exits.

The guards on the upper level proved the toughest nuts to crack, but Marcus had a plan for them. He fired flash-bang grenades from the launcher attached to the underbelly of his rifle. Besides creating a loud noise, the flash-bangs emitted a luminous chaff that was both blinding to the naked eye and capable of overloading the IPGs worn by the soldiers.

The effects of the explosives were painful and disorienting, flushing the entrenched soldiers from their hiding holes.

Befuddled and shaken, the guards clumsily stumbled from cover, clawing at their faces to remove the malfunctioning goggles, giving Marcus the opportunity to pick the defenseless guards off one by one.

Barricaded inside the security room on the upper level, several technicians observed in horror the carnage taking place outside the reinforced window.

The technicians huddled together, watching helplessly as the rapacious attacker decimated their coworkers outside. During the commotion, one of the technicians activated the massive metal security door, sealing off the entire storage area.

Some even clawed at the upper-level security-room door trying desperately to enter, but the technicians inside were so paralyzed with fear, that no one dared open the door. One of the techs inside tried to activate the comm system to report the attack, receiving only static in return.

Fearing that it was just a matter of time before they were next, all they could do was huddle together, hoping that help would somehow come.

Skye warned that if they sealed the storage area, the only way in would be to breach the system inside the upper-level security room. However, entering that room wouldn't be easy since the technicians cowering inside hit the panic button, sealing themselves inside.

Marcus angrily paced the floor.

"Raven Two, come in," shouted Marcus multiple times into his comm unit, but there was no reply. Now he was pissed, vowing to wring Daren's neck when he saw him. Just then, his wayward partner showed up—late, as usual.

"You called?" Daren asked, joking as if they weren't against the clock.

But Marcus was in no mood for games. "You get lost or somethin'?" Marcus shot back.

"How could I?" Daren replied, looking at the aftermath of Hurricane Marcus. "I just followed the trail of bodies."

Marcus started to jump on Daren, but the arrival of the rest of Raven Squad saved him. Marcus took a deep breath to gather himself then addressed the team.

"We need to get that emergency door open," Marcus said. "Skye, you're with me. Tony, Max? You're with Daren."

As Marcus and Skye proceeded upstairs, Daren, Tony and Max took positions near the blast door as they awaited Skye to work her magic. When they reached the top, Marcus, using his tactical glasses, scanned the door looking for any opening. The only vulnerability he could find was a small crack under the door. Perfect, he thought.

"Time to use your new toy," said Marcus to Skye.

"With pleasure," she responded. Skye sat her equipment pack on the ground, reached in and retrieved two marble-sized chrome spheres and placed them on the ground. With a tap of the metallic gadgets, six tiny legs extended from the spheres, changing their appearance to that of small spider-like insects.

She pulled out a touch-screen control module, maneuvering the spiders toward the crack at the bottom of the door. As they neared the room, the bodies of the spiders morphed to squeeze through the tiny crack, returning to their normal size on the other side.

The technicians were so fixated on the door, they didn't notice the small metal insects crawling toward them. Once in position, the legs of the spiders retracted, releasing Marcus' favorite nerve gas from small vents embedded in each device. In an instant, everyone in the room dropped like a sack of rocks.

Skye smiled as the spiders signaled that all targets were neutralized.

"The deed is done," said Skye.

"Alright, do what you do."

Skye pulled an electronic scyther from her pack, interfacing it with the door's maintenance panel.

Within seconds, she bypassed the locking mechanism and the door opened. Pulling up their balaclavas, Marcus and Skye entered the room one after another, looking down the sights of their rifles, ready to drop anything still moving. But as the spiders reported, all targets were down.

"Clear," Marcus called to his partner, prompting them both to lower their weapons.

Marcus examined the control panel, overwhelmed by the dizzying array of buttons, switches, and screens responsible for the facility's security.

He turned to his comrade, relieved that Skye, a genuine scyther, was the one tasked with hacking into the complex system.

Previously known as hackers, the scything community adopted the term when a prominent ISL lawyer said of a master hacker during a public trial, "A computer to a hacker is like a scythe to a grim reaper."

Scythers throughout the galaxy embraced the term like a badge of honor, and Skye was no exception. She was proud of her abilities and worked tirelessly at her craft.

"Make it fast," said Marcus.

After a cursory glance at the security system, Skye released the emergency locking mechanism that controlled the outside blast-proof door.

The system protocols were a lot different from when she was last stationed there, but the firewalls shutting her out were no match for her skills on the holographic keyboard, emanating a few inches from its silvery surface. In less than two minutes, Skye opened the outside blast door. *Not bad, if I say so myself,* she thought while cracking a slight smile.

#

Downstairs at the blast door, Daren and his small team heard the releasing of the locks. Slowly the doors opened, prompting them to move their rifles to the ready. Once fully opened, the team filed in one after the other, with each team member covering the other as they advanced. It was a drill practiced so many times before, the almost rhythmic movements were second nature to them. Before venturing too deeply into the enormous ore storage area, Daren spotted the RAT in the corner, guarding crates that looked different than the rest. He halted his team before moving within the scan radius of the RAT's short-range bio-detector.

"And there's our RAT," Daren said to himself. He then contacted Skye and Marcus on his comm unit. "You guys seein' this?"

#

Inside the security booth, Marcus and Skye could see the threat displayed on the monitors. Scanning the room for additional life signs, other than the team, Skye found none. She then implemented a plan to disable the RAT by sending it into maintenance mode, which, according to her calculations, would buy them enough time to complete their mission.

"Already on it," Marcus responded to Daren over the comm unit. "Do somethin' about their RAT problem," he said to Skye. But as usual, she was way ahead of him. He watched in amazement as her fingers glided over the holographic keys with relative ease. He could tell this was Skye's domain, and he was glad she was on their side.

"Done. Wake-up in thirty," Skye said to Marcus.

"Two, Leader. You got thirty minutes before the RAT wants cheese," Marcus called to Daren over the comm.

In the control room, Skye performed a remote scan to verify the contents of the container over which Daren was salivating.

"The scan is confirmed. It's refined Krillium alright," she told Marcus. "But there's no way they'd leave somethin' like that layin' around."

The entire situation confused Skye. In her experience, the protocols for storing refined ore were the same from station to station. All refined ingots were stored in a secure vault in an adjoining building next to the ore-processing center. Either the administration had gotten sloppy over the years, or this was a trap. After all, no self-respecting thief would pass up such a treasure trove, so this would be the perfect snare.

Marcus shared the same sentiment and responded to Daren. "Negative, Two. Secure target crates only. Do you copy?" There was a delay in Daren's response. Marcus knew what that meant, so he called to him again. "I repeat, Target crates only. Do you read me?" Again, no response. This time he was sure Daren was about to do something stupid, then the words he dreaded came over the comm.

"Sorry, Boss Man, you'll thank me for this later," Daren replied.

"I said stick to the—" But before Marcus could finish his statement, Daren ended the comm session. Marcus was furious. He and Skye exchanged frustrated looks. He wanted nothing more than to march down to that storage room and break Daren's neck, but he had to watch Skye's back as she wrapped up her business in the security room. Any misstep and the entire station's security would go live, bringing an abrupt and violent end to their mission. Though exasperated by Daren's actions, Marcus knew he had to stay put.

#

Inside the storage area, Tony and Max heard the exchange between their first and second in command. Tony approached Daren. "C'mon, D. You heard what he said."

Already feeling uneasy about the mission, Max threw in her two cents. "Yeah, we need to get goin' before that thing wakes up," she said while motioning toward the RAT. "And as much as I'd love to blow the top on that thing, every guard in this station would be here in one half less than no time."

They bombarded Daren with just how many ways this was a bad idea, but Daren wasn't having it, so he pulled rank.

"Hey, it's not up for discussion. I'm in charge here." Daren would let no one talk him out of adding the refined ore to their stash. "You know how much we can get for this stuff?" he reasoned. "Now load 'em up and let's go."

Tony and Max couldn't deny the value of the refined Krillium ore, knowing there were planetary governments willing to pay anything, even if it meant turning to the shadow market to ensure their stockpiles would never run dry. After all, Krillium deposits were becoming rare because of what was known throughout the galaxy as The Terraformer Energy Crisis.

Planetary Terraforming Modules (PTMs) like the ones found on Titan, were massive technological wonders, placed throughout a celestial body's surface. These oversized devices were used to modify the atmospheres and oceans upon otherwise uninhabitable worlds, creating environmental conditions suitable for life.

But all PTMs were on the verge of failing, jeopardizing the lives of trillions across The Milky Way. The reason for that impending catastrophe was because the PTM's primary power source, Krillium ore—found only on certain asteroids, planets, and moons—had been mined to almost non-existence. So, the race to avert the mass extinction of countless worlds had given rise to the shadow-market ore trade, in which refined Krillium was the hottest commodity.

Daren was, in a word, bullheaded. Tony and Max knew there was no talking Daren down. So, in the interest of time, and not getting their collective heads blown off by the metallic sleeping giant in the corner, they complied with Daren's order. Tony went to work, quickly affixing magnetic anti-gravity rods to each side of the crates, causing them to levitate above the ground. He then attached magnetic couplings between the heavy metal containers, creating a train of floating containers. But as they disturbed the first of the refined ore crates to add to their plunder, an unseen circular device beneath one container vibrated.

Daren had a confident smirk on his face, feeling that Marcus would understand once he explained everything. *We're actually doing it*, Daren thought as he led the train of crates being pushed by his companions. But before he could finish running the victory lap in his mind, the RAT finished rebooting itself, waking up early.

"Reset complete. Weapon systems online," the RAT belted out with its deep robotic voice. The walking tank stood up on its two legs and scanned the room with its bio-detector. "Warning. Threat detected," the machine continued. The RAT turned toward Daren and his team. The only reason they weren't already dead is because the RAT's weapons were not fully charged. But that didn't stop the metal monster from charging toward its newly acquired targets.

The sudden activity of the RAT took Daren and his team by surprise. They needed no run order, as the three of them were on the same page. They ran as fast as they could while lugging the train of crates behind them.

At that moment, the RAT opened fire.

CHAPTER FOUR

In an underground bunker, several hundred meters outside the ore storage area, an eight-man ISG Special Forces Unit known as Saber Team stood huddled around a small sensor station. The team leader—Lieutenant Donald Shepard, a stocky ebony-skinned mountain of muscle—gazed at the screen, when his second in command, First Sergeant Alex Chavez, reported in.

"We have movement on the refined Krillium," Alex said, watching the red dot flashing on his screen.

It took every bit of Donald's rigorous training and discipline to suppress the cartwheels he was turning inside. For years he had been hunting the mastermind behind several high-profile terrorist activities within ISL space. Donald spent much of his career hunting criminals. He'd seen it all. But his current targets were true professionals. They had to be for their holo-files to come across his desk.

ISG high command tasked Donald and his renowned Saber Team with pursuing the most elusive offenders operating inside ISL space. It was only in the last year he learned the identity of the team's leader, Marcus La'Dek. All he had to go on was a holo-snap, or holographic snapshot, provided by a high-level snitch he put the screws to a few months back. It wasn't much, but he'd operated with less intel in the past, so he was determined to make it work.

It had to be them, Donald thought. The modus operandi seemed to fit. His gut was telling him that this felt like Marcus as he thought back to the other unsolved crimes, he now attributed to La'Dek and his team. But he wouldn't rest until he had visual confirmation. Who else would have the stones to hit a mining operation so close to his beloved Earth? Plus, the little trick with the security interference was, in his mind, a dead giveaway. His peers thought he was crazy, but Donald had a knack for spotting patterns, even those as obscure as the electronic trickery being carried out on Mining Outpost Alpha.

Expecting this move, Donald compensated by hard-wiring one of the crates with a pressure sensitive silent alarm. A shielded underground wire ran from the device to the concealed bunker, protected from the low-level pulse that was bombarding surface-level electronics. It was a crude system by 32nd-century standards, but it was effective.

Now Donald would see if the months he had spent shaking down every high-level outer-core thug he could get his hands on would finally pay some much-needed dividends. Capturing Marcus would be his legacy, not to mention it would fast-track a career to which he devoted his entire adult life; a career that was costing him both his marriage and a relationship with his son. He kept telling himself it was for the greater good, though his wife of seven years, Rachael, would argue otherwise. *This guy better be worth it*, Donald thought. He shifted his wandering mind back to the task at hand.

Donald moved to the center of the bunker and called his team to stand in formation before him. They moved with a purpose. There was nothing any of those soldiers wouldn't do for Donald, no order they wouldn't follow, even if it cost them their lives. Donald knew this and vowed never to order them to do anything he wouldn't do himself. So, against his superior's advice, the lieutenant preferred to fight alongside his men. He wasn't content with sitting in the rear on a plush sofa, calling the shots as so many of his fellow officers did.

Donald retrieved a black holo-disk from his pocket and activated the device. Above the disk, a holographic image of Marcus on the right and a fuzzier image of Teric on the left appeared. Marcus was clean-shaven and in remarkable shape. He had a light brown complexion, low-cut hair and a tattoo of a raven's talon on his right forearm. The holo-disk's artificial intelligence estimated Marcus to be a little over six feet, two inches tall, weighing in at an even two hundred pounds.

The image of Teric was barely recognizable. But from what Donald could tell, Teric appeared to be a clean-cut, bearded individual with silver-and-black hair in his early sixties. However, there was little in the way of data available on Teric, as even the holo-emitter could not estimate his physical statistics.

Donald's team stood at attention, feet together, backs straight, arms planted firmly by their sides, and eyes front.

"At ease," Donald called to his unit, prompting them to move their feet shoulder width apart, with hands clasped behind their back.

"Listen up, we need Marcus La'Dek alive," Donald continued, pointing toward the image of their target. "We get him, we get Orion's Shield and their leader, Teric Winters," Donald said in his deep voice. He wasn't into giving drawn-out speeches, filled with extravagant and lifeless words. They had a job to do, and everyone in the room was eager to get to work. "Let's move," he finished, and with that, the team broke formation, grabbing their gear as they prepared to move in on the ore pilferers.

#

Inside the mining facility, the mission was going downhill in a hurry as the RAT pinned Marcus and his team in the administrative section. When Skye sent the tank into maintenance mode, its smart targeting mechanism malfunctioned, preventing the RAT from distinguishing between friendly and unfriendly targets. This unexpected glitch benefited Raven Squad, as every remaining guard that converged on their position was cut down by friendly fire.

Marcus and his crew took cover behind the line of ore crates that Daren and his team had dragged into the area. Housed within indestructible Mk17 storage containers, those crates were the only thing in the room still intact as the Gatling guns of the RAT had destroyed everything else.

"You had to be greedy," Marcus yelled to Daren. But Daren had no response, so Marcus turned to Max. "Tell me you tagged that thing."

"Yes... and no," Max replied.

Before they left the ore storage area, Max rigged the RAT with three of her latest creations, a special explosive intended to encase the target in a sphere of pure kinetic energy, designed to contain the blast. She touted the device as the first bomb that could be "safely" detonated indoors. She called it the Kinetic Energy bomb, or K-bomb for short. But she had been having trouble with the detonator, a problem that she thought she had resolved prior to their arrival on Titan.

"What the hell is that supposed to mean?" Marcus yelled to his explosive's expert.

"What Professor Wesner over there neglected to mention was that the friggin' detonator doesn't even work," shouted Daren, trying to deflect the heat he was catching from Marcus.

"Now hold on a minute, honey," Max fired back amid fiery rounds whizzing above their heads. "My tech always works," she said while looking down at her detonator, baffled as to why it wasn't responding. "It's just a little buggy, that's all."

"You'd better fix it in a hurry," barked Marcus. Everything that could go wrong was happening all at once. It was like amateur hour and at the worst possible time. He had to come up with something fast.

"Switch to acid rounds," Marcus ordered. He hoped the corrosive projectiles would penetrate the thick metallic hide of the RAT and corrode it from the inside out. But it was obvious after the first volley, the walking tank's armor was resistant to the specialized ammunition.

"Acid ain't workin'," said Skye, stating the obvious.

"I know, Skye," Marcus yelled in frustration. But before Skye could tell him to watch his tone, which he knew she'd say, even in the middle of a firefight, Marcus turned to Max. "C'mon, Max."

At that moment, Max solved the problem with her detonator. "Got it," she screamed while pressing the button on her device.

A spectacular blue flash occurred, followed by a loud screeching noise. A bluish white sphere surrounded the RAT, causing it to vibrate uncontrollably. Then, in a violent blast of sheer energy, the RAT came apart like a child deconstructing a toy. But the force of the blast was so great that it took out the RAT and half the wall.

The concussive force knocked the crates and the team off their feet. It was a sight to see. Though flat on her back with the rest of the team, Max laughed in between coughs, as she looked at the disaster caused by her creations. But she couldn't help but wonder if perhaps one k-bomb would have done the trick.

"Sorry, kiddies," Max said to the squad. "Went a little overboard."

The team picked themselves up from the ground and gazed upon the gaping hole in the wall.

"Really?" said Skye, still struggling to catch her breath. "I hadn't noticed."

"Grab the crates," yelled Marcus.

The entire crew took up positions near the crates. Relieved that the anti-grav rods weren't damaged during the attack, Tony activated the devices, causing the crates to once again float above the ground. Marcus ordered his crew to exit the building via the newly created hole in the wall, courtesy of Max's K-Bombs.

CHAPTER FIVE

Marcus and his crew departed the facility, arguing all the way.

"Nice, Max. But ya nearly leveled the place," said Tony, awed with the destruction.

"Y'all quit complainin'. You're still alive, ain't ya?" responded Max.

Though there was plenty of blame to go around for the miserable performance of Raven Squad that night, Marcus was most upset with Daren.

"What was that about?" Marcus yelled at his second in command. "I said, Target crates only."

"I was doin' us a favor. That stuff was worth—"

"I don't wanna hear it, D."

Shots rang out, narrowly missing Marcus and Daren's head. Tony deactivated the anti-grav rods to provide a little more cover. The team dug in and returned fire. Just minutes into the engagement and Marcus could tell these weren't the timid guards they faced inside the facility. The way the aggressors moved and communicated led him to believe they were special-ops soldiers. In doubt that they'd even make it to their rendezvous point, Marcus contacted Jason.

"Six, do you copy?" shouted Marcus into his comm unit. "It's gettin' real out here. Need immediate evac."

A short time later, Jason responded over the comm. "Copy that. Raven Six, inbound."

Marcus knew it wouldn't take long for Jason to get the ship airborne, but the minutes felt like hours. To make matters worse, they were running out of ammo. No one expected this level of resistance. Using hand signals, Marcus directed Tony to lie down suppressive fire with his XR-19 Heavy Repeater. Tony took cover behind a section of the destroyed wall and let loose with his weapon. The constant barrage of golden-colored energy bolts from his rifle forced Saber Team to scramble for cover.

Marcus hoped that his team could hold out long enough for Jason to arrive with close air support and a much-needed extraction. He pulled out a pair of electronic binoculars to get a better look. After a quick adjustment of the zoom, Marcus could see exactly with whom they were dealing, and he wasn't happy. Marcus had known of Donald for quite some time. He had a close encounter with the guy about a year earlier during a solo operation.

Donald had no idea how close he had been to nabbing Marcus that night. So, Marcus had much respect for the resourcefulness of that talented soldier. He lowered the binoculars, shaking his head in disbelief. "Great," he said. Marcus activated his comm unit, reaching out to Jason again. "Any day now, Six."

The Indicator thundered onto the scene, firing its dual nose cannons toward Raven Squad's attackers, forcing them to remain in cover. But unbeknownst to Jason, Donald sent one of his ace engineers inside to force the mining facility's air defenses back online, and his timing was impeccable. After squeezing off a few more shots toward Saber Team, the anti-aircraft batteries along the perimeter walls went live, immediately locking onto *The Indicator*. The massive cannons lit up the sky with their crimson-colored fury.

"Oh, crap," said Jason after picking up the incoming fire. He started a series of death-defying evasive maneuvers. Jason pitched, yawed and rolled his way out of trouble, escaping with only a few grazes to the ship's hull.

The intensity of the incoming fire was so great, Jason couldn't get a solid lock on the cannons, forcing him to back off from the facility, beyond the range of *The Indicator*'s cannons. And firing missiles would be useless since they'd be easy targets for the station's missile defense system. He was effectively out of the fight.

"Leader," Jason radioed to Marcus. "The facility's air defenses are active. You'll have to make your way to Bravo. How copy?"

"Solid copy, Six. Leader, out," Marcus replied. He looked at his crew, then at their assailants and knew what he had to do. His mentor, Jax, told him awhile back that if he always took care of his crew, he'd never have to worry about mutiny. And that's what Marcus did since forming Raven

Squad, he took care of his crew. He wasn't about to let his entire team go down when he knew Donald was after him only. If he could just buy them enough time to rendezvous with Jason, Marcus was sure he could give Donald the slip on his own.

"We're movin' to Bravo," Marcus yelled to his team. "I'll cover you guys." He then turned to Tony. "Give me that thing," he said, motioning toward Tony's weapon.

Daren knew Marcus all too well, and he knew what his longtime friend was saying. His first instinct was to tell Marcus that he was crazy and that they were all leaving together. But he was so close to completing the only mission that mattered to him, the mission that would earn him the money and respect he rightfully deserved. Besides, Marcus was the most talented mercenary he knew. If anyone could make it out on his own, it was Marcus. So, he was fine with moving on without their leader.

"Sounds good, let's move," Daren blurted out.

The team looked at Daren as if he were crazy for being so quick to leave Marcus behind.

Skye ceased firing and moved to Marcus' position. She gazed into his eyes. Skye wasn't a crier but tried in vain to suppress the tears welling inside. But at that moment, she didn't care. She would gladly give her life for him and wasn't content with leaving Marcus to the wolves firing on their team in the distance.

"There's no way I'm leavin' you," said Skye.

But there was no time for sentimentality. He didn't want Skye making this decision any harder than it was. So, Marcus interrupted her and addressed the team.

"They're not here for you," Marcus shouted to his crew. "Now move."

As the team prepared to fall back, Skye remained by his side, ready to go out in a blaze of glory. But Marcus grabbed her by the shoulders while trying to contain his own emotions. "Go, Skye," he shouted. Then he noticed the tears rolling down her cheeks and wiped them from her eyes. "That's an order," he said softly.

Skye saw something in his eyes she had never seen before. She couldn't put her finger on it, but she was certain there was no turning him from this path. So, she backed away as Tony moved in to give Marcus his weapon. Tony grabbed Skye by the arm, pulling her away from the fight. Marcus turned toward Saber Team and prepared to fire. As Skye retreated with the rest of the crew, her sadness turned to rage as she glared at Daren.

Marcus stared downrange toward Donald and his advancing team.

"You want me? I'm right here," he yelled to Donald. Moments later Marcus opened up on Saber Team with Tony's repeater.

Forget the rules of engagement. Marcus did his best to cut every one of them down. *It was supposed to be in and out*, he thought. He struggled to suppress the hatred he was feeling for Daren for putting him into this situation. He forced himself to calm, as his rage affected his aim, causing many of his shots to miss wide. Though unable to hit Donald or his team, he pinned them down long enough for Raven Squad to make their escape.

Donald watched Marcus from behind the cover of the outer buildings. He prepared to have his team flank their target, when the sniper he sent to find a high perch outside of the mining complex finally reported in.

"Watcher Two, in position," the sniper radioed in a calm, even tone.

"Report," Donald replied.

"I have eyes on a group of mercs heading away from Alpha, dragging what appears to be the stolen crates. Orders?" the sniper asked with the crosshairs of his scope aimed at Daren's back.

Donald had only a short time to decide. He would have loved nothing more than to have his sniper take down Marcus' entire team, just to show his resolve, but he had orders to follow. His boss, Brigadier General Thomas Kirkland, had clarified that Marcus La'Dek was the sole target on this mission. He wasn't concerned with a bunch of low-level ore thieves. Kirkland wanted Teric Winters, and his intelligence officers had told him that Marcus was his best shot. Like Donald, Kirkland too was eying promotion. He desired ascension to the title of Supreme Commander of the Interstellar Guard. So, the general's instructions were to be executed without the slightest deviation.

"Negative, Watcher Two," Donald replied over comms. "Bring only the subject down," he finished, relieved that he could bring a swift end to this tedious chore.

The sniper moved the scope of his rifle from Raven Squad to Marcus. Sure, he could have fired an ordinary shock round, but Donald wanted to send a message, so a standard-round fired from the rifle of one of the preeminent snipers in the entire Saber Brigade was the order of the day.

Watcher Two was a little disappointed as he locked onto his stationary target. He wanted more movement out of Marcus so he could claim bragging rights among the few sharpshooters on his level. But like the rest of the team, he was sick of Titan and its unforgiving surface conditions. He wanted to go home, so he centered his crosshairs on Marcus' left shoulder, hoping to produce a clean shot, while missing vital arteries. Providing he survived the bullet, it would take his target months of rehab to regain full use of the arm; but at least he wouldn't be dead, which would really suck, Watcher Two thought.

And with one perfectly aimed shot that could be heard for miles, the sniper let loose a round that passed cleanly through Marcus' shoulder, sending him tumbling to the ground. Watcher Two smiled as the cannon that Marcus used against his colleagues was, at last, silenced.

#

Skye turned upon hearing the shot. Before descending a steep hill leading away from Outpost Alpha, she witnessed Marcus stumbling to the ground.

"Marcus," she screamed, while starting to run toward her fallen companion. But Daren grabbed her by the arm.

"There's nothin' we can do," he shouted. "He wanted us to—"

Before Daren could finish, Skye wrenched her arm free of his grasp and nailed him with a left hook that nearly took his head off.

He didn't respond. All Daren could do was look away while wiping blood from the corner of his mouth.

"I hope it was worth it," Skye snarled at Daren.

Skye took one last look toward Marcus, then moved to the rest of the team who were all staring in shock at what happened to their friend and leader. Max put her arm around Skye as they both wept for Marcus. Even Tony, who seldom showed any emotion, was saddened for his fellow warrior. He put his hand on Skye's shoulder.

"We have to keep moving," Tony said. "For Marcus."

They agreed and pressed forward, continuing their trek toward rendezvous point bravo, where Jason was no doubt awaiting their arrival.

But Daren stayed behind for a while, watching as Saber Team descended upon Marcus like a pack of relentless predators, swarming their helpless prey. It was all his fault, and he dreaded facing the crew back at the ship. He watched his brother in the distance for a few moments, closed his eyes, then turned away. *I messed up*, were the words that kept echoing in his mind as he ran to catch up with the team.

CHAPTER SIX

On a distant rocky hill, *The Indicator* awaited its approaching passengers. Jason was hard at work repairing the damage from the recent skirmish. He wasn't quite the skilled mechanic as Tony, but he could handle most minor repairs well enough. Sparks continued to fly as he stood beneath *The Indicator*'s underbelly with a protective mask on, welding a marred section of the ship's hull.

Minutes later, the team arrived carting a train of beat-up crates. Jason turned to see his teammates looking as worn as the cargo they were lugging. He halted his work and ran to meet them.

"Now that was crazy," Jason said, as if still struggling to come down from his adrenaline rush. "It wasn't the mother lode we were expectin', but at least we have..." Jason trailed off as he noticed the dark vibe coming from the team.

Max, with tears still in her eyes, assisted Tony with loading the crates via the ship's aft cargo ramp. Jason then turned to see Skye marching toward him with Daren moving slowly behind.

"Okay. Am I missin' something here?" Jason asked. He looked down the muddy path upon which his friends had just traversed when he noticed the problem. "And where's Marcus?" he demanded. But there was no response. Skye stormed past him without saying a word. Jason knew that look on his sister's face, so he turned to Daren for answers.

"What's goin' on, man?" Jason asked, but Daren responded with only icy silence. "What happened to Marcus?" he shouted. But one look in Daren's eyes and Jason knew their leader wasn't coming.

Daren spoke before the hard inquiry began.

"Start her up, Jay," said Daren in a somber tone. "We need to move."

The thought of leaving without Marcus knocked the wind out of Jason. But they had no choice. All he could do was obey the orders of Raven Squad's new leader, Daren West.

After loading their ill-gotten cargo, Jason launched *The Indicator* from the surface of Titan, punching through the uninviting clouds above. As the ship rocketed toward space, there were no sounds of laughter or tales of heroics shared among the crew. Only the soft hum of the engines remained as everyone, except Jason, who lingered on the bridge, had retreated to their individual living quarters.

Inside his room, Daren sat on the edge of his bunk agonizing over the fate of Marcus. Was he dead? Was he locked away in the bowels of some ship being tortured? He did his best to dismiss the thoughts racing through his mind. With shaking hands, Daren reached within the makeshift compartment near his bunk, from which he grabbed a small canister of zeth he kept hidden from prying eyes. It was a cylindrical device filled with an inhalant delivered through a thin tube, connected to a small nose mask. With the press of a button, he breathed deeply, allowing the narcotic's vapors to travel through the mask, into his nose and straight to his brain. Daren then activated his small rectangular-shaped holo-player, causing it to float in the air, projecting a holographic playlist above its smooth surface.

He selected a track, leaned back on his bunk, and closed his eyes. As the soft tunes emanated from his holographic music player, Daren realized that nothing—not even his precious zeth—could dampen the torment within.

PART TWO

Runner's End

"Welcome... to The End."

— Ian Trent

CHAPTER SEVEN

As Marcus lay on the ground, clutching his bleeding shoulder, he began to think, *Damn, that hurt.* He tried to move, but excruciating pain radiated throughout his body, forcing him on his back, with his eyes shut tight.

One thing about Titan, there were no delicate green meadows or soft patches of rich soil to be found. No beautiful oases beckoning somnolent travelers to rest their weary heads. The moon's surface was, in a word, unforgiving. And were it not for the mud-covered grounds breaking his fall, Marcus would have added a concussion to his list of injuries.

Upon opening his eyes, Marcus found himself encircled by a group of Saber Team shadow tech soldiers, with rifles carefully trained on his head. Fully armored from head to toe in trademark black combat uniforms, no traces of the soldier's identities were visible, as they wore armor-plated masks completely veiling their faces. In fact, the only visible feature piercing their shadowy countenance was the faint red glow of their advanced targeting optics.

Moments later, Lieutenant Donald Shepard broke through the crowd, followed behind by a levitating Automated Restraint Craft (ARC). Approximately three by two feet in size, the ARC was a floating, artificially intelligent support drone, equipped with anti-gravity restraint cables, capable of detaining targets many times its own size.

Donald signaled for the craft to bring the prisoner upright, prompting the ARC to fire a thick restraint cable to the ground from the turret attached to its underbelly. Like a massive boa constrictor, the cable slithered toward Marcus, coiling itself around him, pinning his arms to his sides. The craft initiated the cable's anti-gravity field, lifting Marcus from the ground as if he had no weight at all.

As the ARC maneuvered him face to face with Donald, Marcus groaned, as the machine was not gentle in its approach. A Saber Team medic moved in to treat his shoulder, followed by Donald who stared at Marcus, pleased that his moment had finally arrived.

"You had to know this day would come," said Donald.

Despite his injuries, Marcus still wanted to fight, even though death would've been assured. But all he could see in his mind was the look on Skye's face, just before she and the rest of the team pulled out. He knew that if there was any chance of seeing his crew again, getting himself killed was definitely the wrong answer.

"Nothing to say?" asked Donald.

But Marcus' look said it all. The lieutenant could see the pride in his face and the insolence in his eyes. *But we have ways of dealing with that,* Donald thought. He turned to the medic. "Stabilize him, then finish up at the ship." He then turned to the rest of Saber Team. "You did well, people. Let's go home."

As the ARC carried him away from Mining Outpost Alpha, Marcus couldn't help but wonder if this truly was the end of the road. For years, he had thought of himself as untouchable, as if death itself would come only at a time and place of his choosing. Then he reflected on the words of his mentor and second father Jax.

"Eventually, this life catches up to everyone." Then Jax would say, "The hard part is knowin' when to walk away."

Marcus dropped his head in shame, because he had not heeded those words. He had drawn from the well one too many times. And now it would seem "the life" had finally caught up with him.

#

The ride on the musty-smelling Prisoner Transport left Marcus with a great deal of time to think. He knew the itinerary, days of endless interrogations, followed by a show trial, after which he would likely find himself staring down the wrong end of a very long prison sentence, or worse.

Following a month-long stay at the Luna One Hospital on Earth's moon, ISG Military Police prepped Marcus for the ten-hour ride to Earth.

They didn't even secure a fast ship, which would have made the trip in less than half the time. But Marcus relished the calm before the storm that was indeed to come, so he sat quietly, lost in thought.

"Son, always remember, 'Lost time is never found again.'" These were the words spoken to Marcus by an old lady he had met in some dusty, rundown Interstellar Space Port a few years back. She had whispered the ancient Earth quote as he, in a rare moment of charity, assisted the feeble woman with her bags as she boarded a passenger shuttle headed to the Borelon Cluster.

Marcus wasn't even sure why he had helped. After all, he had thought of himself as a thug, "a pirate" of the highest caliber, albeit one without his own ship at the time. But old habits die hard, because there had still been something inside that wouldn't allow him to stand by and watch her struggle.

At first, he had dismissed the quote as the ramblings of an old woman, entering the early stages of senility. But now as Marcus sat gazing through the porthole of an earthbound prisoner transport, the idea of "lost time" began to take on a whole new meaning.

CHAPTER EIGHT

Hours later, the prisoner transport entered Earth's atmosphere, jarring Marcus awake from his short nap. When he opened his eyes, he could see it hovering proudly in the distance, the seat of power for The Interstellar League of Planets, *Haven City*. Though he'd been to Earth a few times in the past, Marcus had never actually seen the city itself. He had learned in school that Haven represented the pinnacle of human intelligence, engineering and achievement. He was told the city commemorated a time when mankind put aside petty differences and united, reaching toward the stars for the betterment of humanity.

At least that's what they liked to believe, Marcus thought. From his experience, mankind was still the power-hungry, warmongering race it had been throughout the annals of its storied yet bloodied history. But his instructors had one thing right, Haven City, often called "The City of Man" by Xeno races, was indeed a sight to behold.

Because Haven was neither owned nor run by any single nation on the planet, representatives from both Earth and the united planets of the Interstellar League gathered in this central location to help govern the people. Constructed atop a two-hundred-thousand-square-mile anti-gravity platform, Haven City floated above the calm waters of the Northern Atlantic Ocean, between Earth's North American and European Continents. Many of the city's towering, angular-shaped skyscrapers seemed to touch the stars themselves, with legions of personal and public sky-transports crisscrossing the heavens in organized sky-lanes both day and night.

Interspersed amongst the densely packed buildings gracing the city were the military compounds and top-secret Research Labs, identified by their distinctive enclosed bio-domes, some on ground level while others floated in midair, off limits to all but authorized personnel. As Marcus' prisoner transport weaved its way through the congested sky-lanes, the ship approached one of those military bio-domes, Complex 39, a top-secret black site used for interrogating high-value prisoners.

#

Upon reaching ground level of Complex 39, the pace of things quickened. As soon as he exited the ship, the soldiers blindfolded Marcus and dragged him to a dim room furnished with only a small table and two chairs. Moments later, his captor, Lieutenant Donald Shepard, stormed into the room, clothed in full military dress uniform. He sported a chest full of ribbons and medals, representing only a fraction of his accomplishments with the Interstellar Guard's Saber Brigade. Donald took a seat across from Marcus with arms folded, and an icy glare fixed on his prisoner.

Following a momentary stare-down, during which neither man flinched, Donald broke the silence. He activated a small holo-disk he pulled from his inner coat pocket. He placed the disk on the table where it projected the list of charges brought against Marcus.

"Let's see here," said Donald, skimming through the holographic data before him. "You have been busy, haven't you?"

Donald rattled off the list of charges against Marcus, ranging from theft and assault to the attempted murder of government personnel. He then moved into the realm of speculation, trying to pin other acts of terror on Marcus and his crew he believed they committed in the past. Then, with no attempt to mask his irritation, Marcus interposed Donald's speech with a loud sigh of disinterest.

"I'm sorry, you have someplace better to be?" Donald asked. "I could continue, but I'm sure you know where this is going," he said after deactivating the holo-disk. "But there's only one thing we want from you."

Donald leaned back in his chair, which was considerably more comfortable than the metal slab upon which Marcus was sitting.

"Please describe the nature of your relationship with the leader of Orion's Shield, Teric Winters."

"Never heard of the guy."

"Don't play games with me, son," Donald snarled. "You have one chance, and I mean one chance *only* to make things easier on yourself, or I'll—"

"Or what? You'll beat it out of me?" Marcus asked, amused at the notion. "C'mon, Shep."

Donald smiled, amazed at the audacity of his prisoner. He leaned forward in his chair. "I was gonna say, I'll have you dragged to the back and shot."

"I'd like to see you do that," said Marcus. "We both know if you were gonna kill me, you would've done it on Titan."

"I suppose you have a point," said Donald, chuckling at Marcus' futile display of bravado. "And you don't have to talk. At least not to me," he continued. "We have guys for that. And trust me, they are superb at what they do."

Donald stood to his feet and paced the pearl-colored floors of the interrogation room. "Truth is, I did some digging during your stay at the Luna One Hospital. You know, while you were recovering from that little scratch we gave you back on Titan."

"And?" said Marcus.

"I don't believe you're the connection to Teric we're looking for," Donald admitted while making his way back to his chair.

"Sorry to disappoint," said Marcus with a slight grin breaking the corner of his mouth.

"I've always believed Teric's right hand to be someone else. Not some overconfident punk who ain't nearly as good as he thinks he is," said Donald.

And on your first point, you'd be correct, Marcus thought. Besides, the one they wanted was a guy named Samson Kull. But Marcus would leave that to the ISL's *unintelligence* community to figure out.

"If I'm not your guy, then why the face-to-face?"

"Don't get excited, kid. I close out all my operations personally," Donald replied. "If it were up to me—"

"Too bad it ain't up to you," Marcus interrupted with a bullish smirk on his face.

"Yeah, too bad," said Donald.

After another momentary stare-down, the lieutenant stood up and moved toward the door. "We're done here. Enjoy your stay on Earth. I'm sure you'll find the accommodations we've arranged, most befitting a man of your esteem." Donald opened the door, giving way to three large, muscle-bound guards who took Marcus into custody, while sneaking in a few cheap shots.

"See ya at the trial, kid," said Donald as he exited the room.

As the door closed behind him, Donald could hear the sounds of Marcus becoming *acquainted* with his new friends inside. After a quick check-in with his wife, whom he'd seen little of in the last ten months, he continued down the dimly lit hall toward the exit, optimistic that justice would finally be served.

CHAPTER NINE

"Marcus La'Dek, I hereby find you guilty of crimes against the Interstellar League of Planets," said the judge as she slammed her gavel. These were the words spoken to Marcus that concluded two years of intense hearings, lengthy investigations, and a healthy dose of political grandstanding. The ISL was starved of results in its unofficial war against the Orion's Shield terror network.

But how does one fight an enemy that is largely unseen, blending among the masses, spanning many star systems? These were the unspoken questions with which the ISL had wrestled for centuries. And for the last two years while incarcerated on Earth, Marcus became the closest thing the ISL had to a secret weapon to combat their longtime nemesis.

Donald often questioned the true effectiveness of the intelligence provided by Marcus, believing the disgraced mercenary was often holding out on them. But much to the lieutenant's consternation, he witnessed his superiors tripping over themselves time and time again, acting upon every drop of information gleaned from their most valued asset.

But Donald shifted those thoughts from his mind. He wanted nothing spoiling the moment. And now he would see the great Marcus La'Dek locked away for the rest of his natural life. So, he sat in quiet expectation inside the closed-door trial, awaiting the next words from the judge's mouth.

"Mr. La'Dek, you are convicted on all counts, carrying a *32.15 full-cycle* prison sentence, to be served on the penal colonies of Mars," the judge continued.

As she spoke, Donald looked on with an overwhelming sense of pride, thrilled to see the malevolent mercenary on the verge of getting what was

CARGO 3120: TIES THAT BIND

coming to him. He turned his gaze to Marcus, as if searching for a reaction to his newly determined fate.

Marcus had no desire to give Donald the pleasure of such a reaction. But when the judge said his sentence was 32.15 full cycles, he couldn't help but lower his head, knowing that according to the universal galactic time scale, his prison time equated to just over thirty-seven Earth years. In Marcus' mind, his life was over. He glanced to his left where he caught Donald staring at him with a smug, condescending look on his face. It was a look he'd seen time and time again from those who thought themselves to be his better. But in that moment, there was nothing he could do. So, he stood within the courtroom in total silence as he watched his entire world come crashing down before his eyes.

While thirty-seven years wasn't *the long ride* for which Donald had hoped, it still satisfied him that Marcus was given the minimum sentence, as it would effectively take him out of the mercenary game, for good.

"However," the judge continued, bringing a sudden halt to Donald's internal celebration. "Considering the extraordinary cooperation that you provided over the past twenty-four months, which led to our forces dealing a major blow to the foundation of Orion's Shield, it is this court's determination that your sentence be reduced to 6.96 full-cycles, plus time served."

Marcus raised his head, feeling as if new life had returned to his body. He looked at the judge, astonished by this sudden turn of events. As he let out a sigh of relief, he turned to Donald, who looked as if he were moments away from exploding.

"What?" Donald roared inside the silent courtroom.

"Lieutenant Shepard," the judge responded, appalled at the uncharacteristic outburst, "You are out of line. I will not have—"

"Major blow? We barely scratched the surface! We're no closer to Teric Winters than—"

"I said, Enough, Lieutenant," said the judge as she arose from her seat. "You will contain yourself, or I'll have you thrown out of my courtroom."

Marcus couldn't suppress his laughter, seeing the Lieutenant come unhinged in that manner. He locked eyes with Donald, giving him a look of silent victory, which only infuriated the disgruntled soldier even more.

"Bunch of incompetent bureaucrats," Donald continued, forcing himself not to use more colorful terms.

But he took personal offence to the judge's decision, as it was often *his* team that was put into harm's way over the last year and a half, executing operations that felt more like Marcus was settling personal vendettas between him and his one-time rivals than anything else.

"You're so concerned with your public image that—"

"That's it," the judge said. "Get him out of here," she roared to the courtroom guards.

The armed soldiers approached Donald and grabbed him by the arms. But Donald wrenched himself free of their crushing grips.

"I know the way out," he said, making an about-face turn, and storming from the room.

Marcus basked in the moment's glory as he watched his longtime rival exit the courtroom in disgust. It was a small victory to be sure, but a victory nonetheless. If only they knew how right Donald was, he thought. Though the ISL viewed the targets he provided as *high value*, they were ultimately insignificant arms of Teric's infrastructure that could be replaced with little effort. But he had to make it look good. So, Marcus would often send them after criminals with whom he had negative dealings in the past, just to watch them burn. And indeed, many of them did burn at the hands of the merciless shadow tech soldiers of Saber Team, led by their famed commander, Lieutenant Donald Shepard.

While Marcus was relieved to know that he wouldn't spend the better part of his adult life rotting in the notorious work camps of Mars, he still felt eight years, plus the last two he spent locked up on Earth, was still a great deal of time to lose. As the guards escorted him from the courtroom, Marcus once again reflected upon the words of the elderly woman he met on that Star Port all those years ago. And he figured the old lady had it right. When it came to *lost time*, once it's gone, it's gone.

#

Following the trial, they wasted little time shipping Marcus to what would be his home for the next eight years, Martian Penal Colony-2301, in the Cydonia region of Mars. Once dubbed "The Red Planet" because of its iron-oxide-rich regolith, large portions of Mars were more Earth-like in appearance by the year 3111, thanks to a millennium of terraforming, which brought back many of the planet's oceans. The terraforming also brought about a total restoration of the planet's previously unbreathable atmosphere.

Resting upon the site once called *The Face of Mars*, Penal Colony-2301 was one of three hundred state-of-the-art maximum-security prisons from which no prisoner, human or alien had ever escaped. Though officially designated MPC-2301 by the ISL, its inmates knew it by another name, *Runner's End*.

Before first visiting Mars, humanity had written off the original site of Runner's End as little more than a natural geological formation, despite many arguments to the contrary, especially among those who believed that humans were not alone in the galaxy.

But it stunned the scientific community when, upon visiting the planet in the year 2035, they discovered that *The Face of Mars* was the small soil-covered tip of an expansive subterranean way station, built by an ancient, technologically advanced civilization that had long since departed Mars.

Though many theories existed as to the exact purpose of the way station, the ISL government, centuries later, converted the advanced structure into a high-tech prison instead of the laboratory for which many in the scientific community vehemently pleaded. But the prison's crown jewels were its innovative stasis chambers, reverse engineered from existing technology found on site, which was later used as an advanced form of long-term solitary confinement.

Marcus had heard of Runner's End and the horrors described by those who ended up in stasis. They told him it was like a living nightmare, as subjects were placed in cylindrical chambers with diodes placed throughout their bodies, providing the electrical stimulation necessary to prevent muscle atrophy. Intravenous feeding tubes were inserted into the body, ensuring prisoners received the necessary nutrients for the long dormancy period they were soon to face. Finally, a stasis energy field was activated, placing the subject into a long-term state of suspended animation.

During stasis, the prisoner aged as normal and was conscious the entire time. Subjects were more than just prisoners of the facility; they were prisoners within their own bodies. But Marcus' greatest fear was not Runner's End, nor the infamous stasis chambers found within. The thing he feared most was deciding the path he would take after prison. The thought gave him headaches, for the mercenary life was all he knew, and loved.

As Marcus sat amongst the rowdy soon-to-be inmates on the armored hover transport, he closed his eyes, trying to refocus his thoughts, lest he drive himself crazy. An hour later, Marcus opened his eyes to see Runner's End coming into view through the ventilation slits of the armored transport.

When the large bus-like hover-vehicle came to a complete stop, one of the armed ISG soldiers opened the rear hatch. She observed the rambunctious prisoners for a moment, as they seemed to ignore her presence.

Discharging her sidearm into the air, she shouted to the inmates. "Everybody shut up and get out."

In an instant, there was total silence. One by one, the shackled prisoners, including Marcus, filed out of the transport, lining up in a single column. From outside the vehicle, Marcus could see the true scale of Runner's End. The facility looked more like an impregnable fortress than any prison he'd ever seen. Runner's End had it all, including Enhanced Bio-detectors and automated turrets with state-of-the-art target-recognition upgrades. Swarming throughout the prison was an endless contingent of guards, including military-grade Android Security Units with enhanced neural networks, making them as lifelike as the human guards they supported.

Besides the guards patrolling the facility, no less than three active battalions of the walking Robotic Autonomous Tanks roamed the prison grounds, operating around the clock. All the prison's physical security measures interfaced with the enhanced, real-time monitoring systems, based on advanced technology left behind by the ancient alien race that had inhabited the site eons before. There was no security system in the known galaxy like the ones found on the Martian penal colonies; and the systems within Runner's End were light-years ahead of the other 299 prisons on Mars.

Moments later, the warden, Ian Trent, approached the inmates, flanked by seven armed guards. The warden's entourage looked as if they were searching for the slightest reason to use the fresh batch of inmates for target practice. Ian was a tall, husky man with piercing, blue eyes and dark, slicked-back hair, dressed in a black business suit, covered by an ankle-length jacket and black boots. Taking great pride in his appearance, Ian looked almost out of place in the harsh environment surrounding Runner's End.

As the warden and his men halted in front of the convicts, the guard who directed the prisoners out of the transport approached.

"All prisoners accounted for, Mr. Trent," said the guard as she handed Ian a circular holo-disk containing the passenger manifest.

"Excellent, Mrs. Hall," Ian said as he pocketed the disk. He turned to address his new *guests*.

"My name is Ian Trent, director of this fine establishment you see before you. And believe me when I tell you, we have seen it all," Ian continued in his proper, southern-American accent. "Some, or perhaps all of you are thinking, 'I won't be here long,'" he said. "But I implore you, abandon this foolish line of thinking. No one leaves here, unless I decree it," he said, pacing the rust-colored soil.

"We have housed everyone from the treacherous, silver-tongued Lyrian, to the indomitable, yet dimwitted Gorean Cyborg," the warden continued. "Yes, sir, we've had them all. But in the end, everyone complies. No one runs."

Ian pulled a black baton from the holster of one of his bodyguards. With the press of a button on the rod's hilt, the baton pulsated with yellowish-white arcs of energy, dancing around the business end of the weapon like miniature bolts of lightning.

"They call this place Runner's End," Ian continued, pointing to the prison complex behind him. He moved face to face with Marcus, staring into his eyes. "But you will soon come to know it as, *The End*."

Marcus stared back, undaunted.

"Look at what we have here," Ian said to his guards. "A tough guy who thinks himself an exception to *my* rule."

Without warning, Ian jabbed the tip of the baton into Marcus' abdomen, sending an electric current surging through his body, bringing Marcus to his knees.

Enraged, Marcus charged toward Ian despite his shackles, but the Warden's bodyguards intercepted him. The soldiers proceeded to pummel Marcus with their fists and batons, while Ian looked on in approval.

As the guards savagely beat the defiant prisoner, the rest of the inmates watched in silence. After a few moments, Ian raised his hand, prompting his soldiers to cease their assault. Marcus somehow staggered to his feet in total defiance. But the soldier standing behind him brought Marcus to his knees with a single swing of the baton.

Ian stood in front of Marcus, looking down on him. "I'm sure we'll come to know each other quite well."

As the soldiers began marching the inmates toward the ominous black gates of Runner's End, Ian turned to address the convicts one last time. "Welcome... to The End."

CHAPTER TEN

It was customary for certain prisoners, upon arrival to Runner's End, to be assigned caseworkers as part of the ISL's Rehabilitation & Reintegration Program (RRP).

Ian absolutely hated the idea. If the warden had his way, he would have redirected the funds from that overpriced charity fest into additional security upgrades for his prison, or other *off the books* activities. But he had no choice. Every year some bleeding-heart activist on Earth would lobby for the fair treatment of prisoners, demanding the government give inmates the chance to reintegrate into ISL society.

The RRP targeted ISL citizens serving less than twelve years, believing they stood the greatest chance of successful reintegration. And given the ISL's precarious economic situation, they needed every able-bodied citizen injected back into the workforce. So, between the RRP guidelines and the intelligence he provided during his closed-door trial, Marcus earned himself a first-class ticket into the program.

While final preparations were made for his cell, the guards escorted Marcus to a small, modestly furnished, well organized office. In the center of the room was a bench, a large desk and a much more comfortable chair. He knew which of the two was meant for him, so he shuffled his way to the metallic bench whilst trying to keep from falling flat on his face, thanks to the indestructible shackles bound to his wrists and ankles.

As he awaited his caseworker, Marcus sat under the observant mechanical stare of two armed Android Security Units (ASUs). There was something unnerving about the soulless gaze of the ASUs, he thought. Marcus knew the androids, absent all emotion, would cut him down without the slightest hesitation.

Many questioned the Warden's use of such a large synthetic security force. But Marcus figured it made perfect sense. When dealing with the likes of those incarcerated at Runner's End, why would The Warden entrust those susceptible to coercion, intimidation, or compassion to stand watch over his most prized assets? The Warden had a lot to lose, and a great deal more to keep secret from what Marcus had heard of the all-powerful Ian Trent.

Before he could finish contemplating the multitude of conspiracy theories surrounding Runner's End and its corrupt leadership, Marcus' assigned counselor, Charles Harris, stepped into the room.

Charles was an older, balding gentleman, almost a relic from another time. He entered the room with notes written on actual paper, a rarity in those days, as most people opted to use the flexible, paper-thin digital pads, capable of converting voice and even the worst of handwriting into clean, digital text with astounding accuracy.

From his observation, Charles seemed to have a no-nonsense demeanor about him, complete with eyes that seemed capable of peering into the very soul of an individual. Marcus had a knack for reading people, even upon first meeting them. His former teammates called it a gift, but he saw it as being cautious. Hoping to get some burned-out government worker looking for a quick file closure, Marcus was dismayed to find himself staring at the complete opposite.

As Charles made his way to the chair behind his desk, he noticed his client didn't take the more comfortable chair and smiled. He turned to address the two Android Security Units. "Release his restraints," the counselor ordered.

"Sure that's a good idea?" Marcus asked.

Charles shot another soul-piercing glance toward his client. "I'm sure I'll be fine," he replied.

But before Charles could finish his next thought, the two ASUs began to protest in unison. "MPC-2301 Security Protocol 15 states, all prisoners are to remain—"

"Override SEC Protocol 15. Authorization: Theta-eight-nine-seven-four," said Charles.

"Authorization Confirmed," the two guards responded.

After releasing Marcus' bonds, they made an about-face turn and exited the room, posting themselves on either side of the entrance.

"Nice trick. Trained 'em yourself?" Marcus asked.

"In a past professional life, I helped design and program those things," Charles replied. "They, like most people, are harmless, if you know how to talk to them," Charles continued as he began organizing his papers. "I'm Charles Harris. I'll be your ISL Reintegration Case Worker. Pleased to—"

"Programmer, huh?" Marcus asked.

"Yes, sir. Twenty-five years," Charles responded.

"So, you went from reprogramming bots to people, that right?" Marcus asked.

"If only people were that simple," Charles responded, amused by Marcus' attempt at sarcasm. "People are a bit more... complicated."

Charles moved from his chair to the front of his table, upon which he sat in front of Marcus. He stared at him for a moment, then spoke.

"Say you want that droid over there to change its position," Charles said, pointing to the ASUs through the transparent office door. "The process is simple, *if* you know its programming language. But a person, on the other hand, won't change their position until it first makes sense *why* such a change is necessary."

"And if they don't want to change their *position*?" Marcus asked, deciding to play along.

"Perhaps the reasoning as to why they *should* change hasn't been communicated in a language they understand," Charles said.

Marcus could see where Charles was going with this. He wasn't about to let some antiquated programmer turned shrink tell him what to do with his life.

Noticing Marcus' withdrawal from the conversation, Charles changed the subject.

"I'm sure they explained the purpose of this program to you," said Charles. "It's my job to assess and determine if you should be reintegrated into ISL society following your incarceration. So, I urge you to take our sessions seriously, Mr. La'Dek," he continued. "And trust me when I say, I'm the closest thing you have to a friend in this place. But on the bright side, I'd say you're off to an excellent start."

"We done?" Marcus asked.

Charles could see that he would have his work cut out for him. But Marcus' subconscious decision not to take the chief seat in the room spoke volumes. So, he stood to his feet and extended his hand. But to no surprise, Marcus didn't reciprocate the gesture.

Charles then escorted his client to the awaiting ASUs, after which he returned to his desk to work on his case plan. As he continued his duties in total silence, he was sure there was still something within Marcus worth salvaging, even if the battle-hardened mercenary couldn't see it within himself.

CHAPTER ELEVEN

"Mr. Trent's campaign speech the other day was laced with the same inflammatory, xenophobic rhetoric that's been the centerpiece of the Orion's Shield belief system since they were first branded terrorists seven centuries ago," said Nadia Romanov, Secretary of Xeno-Affairs within the Interstellar League's current administration, serving at the pleasure of the president, Jonathan Vance.

The secretary was a tall brunette woman of Russian descent, now living in Earth's Haven City. Adorned in the customary attire for ambassadors and other high-ranking officials, Nadia wore an extravagant white ceremonial robe with ornate patterns of gold and silver, woven into the lapel and seams.

The secretary was the subject of an interview conducted by Bobby Wiseman, the fast-talking, Earth-based reporter from the Galactic News Agency, also called the GNA. And for the past thirty minutes, Nadia had been answering Bobby's questions regarding the upcoming ISL presidential election between Jonathan Vance, who was campaigning for a second term, and his political rival, Ian Trent, the infamous warden of Runner's End.

"Wait a minute," said Bobby. "Are you comparing the Trent campaign to that of a terrorist organization?"

"I'm not merely drawing a comparison," said Nadia. "I'm saying Trent's campaign is both financed *by* and in league *with* Teric Winters himself."

"That's some theory," Bobby responded, stunned by Secretary Romanov's words. "Tell the viewers watching today. What is the basis for these claims?"

"Think about it, Bob. In the last three months, how can one explain the wild turn of events that has caused Mr. Trent to become the *sole* candidate opposing President Vance?" asked Nadia. "Two presidential frontrunners walk away from their campaign bids for so-called 'personal' reasons. Three more mysteriously go missing without a trace," she continued in a stern yet passionate tone. "And let us not forget Garec Solomon, planetary governor of Tralox Seven and father of five, who allegedly committed suicide by shooting himself in the back of the head, four times—"

"I'll be the first to admit, the circumstances are troubling, suspicious even," said Bobby. "But again, what proof do you have that Ian Trent had anything to do with—"

"History, Bobby," Nadia said, expecting the reporter's question. "These were the same schemes used by Zedolf Winters, the founder of Orion's Shield, and father of Teric Winters himself," she said. "Zedolf used these exact tactics to clear the field in their failed attempt to get Teric elected six hundred fifty years ago. And after their stunning defeat, they resorted to bombing the Lyrian Embassy here on Earth, prompting *The Great Purge*, the bloody conflict that drove Orion's Shield to the outer core. And the rest, as you know, is history."

"Careful now. These are serious accusations, Madam Secretary."

"And we do not make these claims lightly," Nadia responded. "Ever since that monster, Teric Winters, unnaturally extended his life through *illegal* genetic manipulation centuries ago, he's been plotting new ways to infiltrate our government," she said. "He can't do it himself because of the many atrocities committed by his hand. So, we believe he's found another means to take power. And his name is *Ian Trent*," Nadia concluded.

Bobby was at a loss for words. He started to speak, but Nadia leaned forward in her chair, as if making an urgent plea to the billions that were no doubt watching the live interview across the galaxy. "Ian Trent is a dangerous man, Mr. Wiseman. And men like *him* simply cannot be handed power, under any circumstances."

#

At Runner's End, Ian watched in anger as Secretary Nadia Romanov continued to eviscerate him during her live interview with Bobby Wiseman. The warden deactivated the eighty-inch view screen in his profligately decorated office, then turned toward his campaign chief and trusted advisor, Malak Pearce. Before Malak could speak, Ian hurled a nearby ancient piece of Sorean pottery into a wall, shattering the beautiful and incredibly rare work of art into thousands of pieces.

Taken aback by Ian's response to the GNA interview, Malak frowned, having never seen the warden act in such a manner.

For once Ian had no words. He turned and gazed into the blank view screen across the room, fuming over the secretary's words.

"They're reaching, Ian. Like you've always said, The ISL's withdrawal from the GPC is long overdue," said Malak, referring to the Galactic Planetary Consortium, the interstellar ruling body of the Milky Way, comprised of the ten prominent races of the known galaxy, including its smallest and least influential member, the human-led ISL faction.

But Ian continued his tantrum, catapulting another piece of priceless artwork across the room.

Malak moved in to calm him down. "Look, your message is resonating with the people," said Malak. "President Vance and his cronies like Nadia can't see it, but it's time for us as humans to stand on our own two feet, without alien interference."

"That idiot, Wiseman, should be fired," said Ian, ignoring his advisor's words. "How could he let Vance's lapdog soil the air with those baseless fantasies?"

"Trust me, we'll be on the air with Wiseman and his colleagues by tomorrow to set the record straight," Malak assured his boss. "Teric Winters? I mean, where would Nadia get such a ludicrous notion?" he pondered while pouring Ian a glass of a rare vintage French wine that he had imported from Earth.

"Our oppression by the Xeno-races of the GPC has only intensified under Vance's watch," said Ian while pacing the floor of his office. "And they have the nerve to take issue with *my* tone?" he continued. "The winds of change are blowing, Mr. Pearce. You mark my words."

"They are indeed. People are fed up with the lack of leadership under the current administration," said Malak as he poured his own cup of wine. "Trust me, mankind will not tolerate another ten years under Vance."

"Ever the optimist," said Ian. "Go, arrange those interviews. It's time we went on the offensive with our message."

Malak nodded in agreement, downed his drink, then exited the office, leaving Ian alone to contemplate his next move.

Malak Pearce was one of the few people Ian respected, having grown up with him on Earth's North American continent in the state of Alabama. But in some ways, Malak was insufferably naïve. He did not understand the sacrifices and compromises one must make to usher in *true* change. But Ian needed a squeaky-clean face on the campaign to help make his message more palatable for the masses, so he allowed Malak to lead that charge.

Campaigning on an interstellar level was a long and arduous process, involving thousands of staffers across thousands of planets, all to help spread the message of their candidate. The process also required talented leaders like Malak Pearce at the top, working to coordinate all the moving pieces while keeping everyone on message.

Though presidents of the Interstellar League served terms lasting ten years, potential candidates looking to challenge the president at the end of their term would hit the campaign trail early. The campaign season always began during the current administration's fourth year in office, allowing the potential candidates a much needed six years to sway the hearts and minds of the ISL citizens, trying to convince the people to elect them to serve as mankind's leader on the galactic stage.

It was President Jonathan Vance's fourth year in office, and Ian Trent was just beginning his campaign efforts, which were already off to a rocky start.

Ian engaged the locking mechanism on his door. He then started a shielded, audio-only transmission, impossible to intercept by those whom the Warden felt were no doubt spying on him. It was a call that he couldn't make in front of his friend, knowing Malak would never understand that sometimes the ends justify the means, no matter how dark and ugly those *means* might be.

"This better be good," said the distorted voice on the other end of the transmission. It was Teric Winters, responding from somewhere deep within the outer core.

"The prisoner you told me to look out for, Marcus La'Dek, has arrived," Ian reported. "And per your request, I extended our new guest a very *warm* welcome."

"Excellent," Teric responded. "Now to conclude the matter. I want that mercenary dealt with."

"With all due respect, Mr. Winters, he looks like a broken man. I don't think he will be a—"

"He betrayed me, Mr. Trent," said Teric. "A personal affront that a man in my position, cannot let abide."

"But, Mr. Winters, he's a high-profile inmate, and we're in the middle of a campaign," Ian reminded Teric. "I can't afford another scandal at my prison right now. It would mean the end of our bid for the supreme seat in the Citadel," he said, approaching the reinforced window of his office overlooking the vast prison grounds below.

"We're already under extreme scrutiny by the Vance administration as it is," said Ian while fidgeting with the empty wine glass in his hand. "That Nadia Romanov is the real threat. She's the one you need to put down."

"Are you giving me orders, Trent?" Teric asked, as he grew tired of Ian's rambling.

"By... by no means, Mr. Winters," Ian said following an unnerving silence, realizing that his reckless words may cost him more than just the election. He swallowed hard, his heart nearly pounding through his chest.

"Don't forget who it was that got you here," said Teric. "I don't care how, I don't care when, but Marcus doesn't leave your prison alive. Do you understand?"

"I live only to serve," Ian said. He was loyal to Teric, and a true believer in everything for which Orion's Shield, humanity's true protectors, stood. Ian wouldn't rest until Teric's vision of a human-dominated galaxy came to pass. "But it must be done in a way that doesn't point back to me."

"But of course," said Teric. "Just get it done." And with that, Teric ended the communication, leaving Ian the unenviable task of doing Teric's bidding while keeping his own hands clean.

Ian dropped the wineglass to the floor and began guzzling straight from the bottle. He returned to his desk where he brought up the holo-file of Marcus La'Dek. As he sat staring at the holographic image of the fallen mercenary, he began to think, *just another means to an end.*

CHAPTER TWELVE

As the chrome Android Security units escorted Marcus down the concrete halls of the maximum-security wing of Runner's End, Marcus began feeling as if all eyes were on him. Populated with mostly human inmates, all prisoners dressed the same, wearing taupe-colored short-sleeved jumpsuits with brown work boots.

The multi-tiered housing unit contained countless rectangular-shaped two-man cells on every floor, each unit furnished with a metallic bunk bed, a single waste-disposal unit, and a sink with no mirror. Made of bluish-white pulsating rods of energy, the cell bars were painful to the touch, but non-lethal to the occupants.

As he made his way to cell forty-seven, located on the third floor, Marcus observed what looked to be an endless contingent of human and android prison guards roaming every level of the 137-story housing section.

Floating ARCs, like the one that restrained Marcus on Titan two years earlier, followed many of the guards. The small levitating support drones conducted bio-scans on the inmates, checking for contraband. Besides the restraint cable turrets found on the underbelly of each ARC, they were also fitted with twin pulse cannons, an odd upgrade for supposedly non-combat support craft.

Several feet in front of Marcus, two armed human guards, dressed in gray and black battle dress uniforms, marched another prisoner toward a different cell on the same floor.

Without warning, the inmate bolted, screaming at the top of his lungs, vowing he'd never return to that cell. It was clear the man wanted to die... and the brutal security forces of Runner's End seemed all too eager to oblige.

"Halt, Inmate two-three-nine-seven-two. Return to your escort," belted the levitating ARC unit flanking the human guards. Its electronic monotone voice boomed across the housing unit, nearly drowning out the racket coming from the rambunctious prisoners throughout the area. But the inmate sprinting down the hall ignored what would be his first and final warning. One of the human guards pressed a few buttons on his holographic wrist-pad.

"Order received," said the ARC unit.

The restraint craft engaged its maneuvering thrusters and moved in on the fleeing prisoner, closing the gap in seconds. Instead of deploying its restraint cables, which would have been standard operating procedure anywhere else, the ARC unit engaged the inmate with its rapid-fire pulse cannons. The orange bolts of energy ripped through the inmate. The projectiles burned so hot, the rounds cauterized the wounds as they pass through the prisoner's flesh, leaving no blood in their wake. The inmate fell to the ground, riddled with holes that were still glowing hot.

They sent a message loud and clear to any other would-be escape artists, *if you run, you die. No exceptions.* Most prisoners knew the rules and would never think of running. But those unfortunate enough to witness the execution retreated farther into their cells in total silence, reminded of the words Warden Trent first spoke to all prisoners, *"No one leaves, except I decree it."*

It was overkill to be sure, but it wasn't anything that Marcus hadn't seen before. He watched as the ARC lifted the dead prisoner from the ground with its restraint cable, carrying him away from the housing unit. Moments later, Marcus arrived at his own cell.

"Put on your jumpsuit and report to the common area," said one of his Android security escorts.

Marcus sighed as he complied with the guard's order. It wouldn't be easy, but he had to accept his new reality at some point. Eight years was a long time. And given the warden's contempt for him upon their first meeting, he expected a rough stay at the most secure facility in the Sol system.

#

The common area was the closest thing inmates had to a place of rest from their labor in the Martian work camps. It was a large room with off-white walls, a metal floor, and bench-style seating of stone. There was also a large exercise area, giving inmates a means of relieving stress, a necessity given the high-stress environment that was Runner's End.

Marcus looked throughout the room, seeing perhaps the single largest gathering of nefarious criminals he'd ever seen congregated under one roof... and that which he saw was only part of the prison population. It was a scene played out time and time again within any penal facility, inmates socializing in various cliques, gangs at odds with each other, trying to assert their dominance within the prison hierarchy of power and control, and, of course, the strong preying on the weak.

Marcus had no desire to mix it up with anyone on his first day, so he found a quiet corner of the room where he hoped to keep to himself until it was time to return to his cell. He wasn't there to make friends. He just wanted to do his time and move on.

Moments after taking his seat, Marcus turned to see a man being brutally assaulted by three heavily tattooed men. The victim was an inmate known only by the name, Jackson. He was a short, scruffy-looking individual of average build. He had shoulder-length purple hair, green eyes, and an unkempt beard. He tried in vain to fight back, and the lack of guard coverage in the room didn't help matters either.

Deep down, Marcus had trouble watching people struggle. He often felt compelled to help. It was a trait instilled in him by his biological father years ago, though he sought to suppress it when he turned to the life of a mercenary. But the trait seemed always to rear its ugly head when he'd rather just play the background. While he had no desire to get involved, Marcus couldn't take hearing Jackson's cries for help as the fists of his attackers collided with his body and face. So, against his better judgment, he moved toward the disturbance.

As Marcus neared the three attackers, he tried convincing himself that helping this guy would be the perfect chance to send a message. He knew it was just a matter of time before some would-be tough guy tested his resolve, being that he was the new guy on the cell block. As Marcus inched ever closer, an assailant going by the name, Sarak, moved to intercept.

"This ain't got nothin' to do with you, man," said the muscle-bound inmate. "I won't warn you again."

"C'mon now. Don't you at least wanna make this interesting?" said Marcus. "Or shall I leave you ladies to it?"

Furious, Sarak attacked. But before he could land a single blow, Marcus delivered a debilitating kick between the legs that sent him to his knees, squealing in pain. Marcus then landed a rising knee strike to the chin, knocking Sarak cold.

The second attacker, a much skinnier man than the first, calling himself Byrd, turned to see Marcus taking out his partner. He rushed in, swinging with a series of unskilled haymakers, which Marcus dodged, blocked, and parried with little effort.

Unable to hit the elusive target, Byrd tried taking the fight to the ground. He grabbed Marcus by the collar, but Marcus countered the move by gripping Byrd's arm with one hand, and grabbing him behind the neck with the other, preventing his opponent from moving. Marcus leapt into the air, hooking one leg around Byrd's neck, using his weight to flip the thin man to the ground onto his back. With perfect form, Marcus transitioned into a double-leg arm-bar, snapping Byrd's arm in two, causing the inmate to cry out in agony.

By now, a crowd of inmates looked on, clamoring for more. The third and final attacker approached. He went by the name Razorback because of the cybernetic spinal column protruding from his skin, resembling jagged spikes. It was a hack job by some scumbag doctor to repair a back injury he sustained years ago during a shootout with ISG forces. He thought it was a cool look, so he never sought to have the procedure corrected.

Razorback ceased his assault on Jackson and turned his sights toward Marcus. Like a feral animal, he rushed in, but Marcus used the man's momentum against him by hooking Razorback's arm with his own, flipping him to the ground with a carefully timed hip throw.

But Razorback sprang to his feet, rushing again. But he found himself on his back as Marcus sidestepped the rush and delivered a devastating clothesline takedown. Before Razorback could move, Marcus pinned him to the ground with his knees, raining down a flurry of punches, hammer-fists and elbow strikes.

As Marcus continued to beat Razorback to a bloody pulp, Jackson staggered to his feet and approached him from behind. Using what little strength he had left, Jackson pulled Marcus off Razorback before he beat the man to death.

"Get down, man. The guards are comin'," said Jackson.

Moments later, the Android Guards swarmed the area, ordering everyone to get down on their faces. After securing the room, the androids faced the inmates.

"Clear the area immediately, or we will open fire," belted one android.

Within minutes, the guards cleared the room except for those involved in the melee. As the rest of the inmates exited the common area, a host of heavily armed guards escorted the prisoners back to their cells. While the guards sent Jackson and his three attackers to the infirmary, an ARC unit restrained Marcus, taking him to a temporary holding cell while prison authorities investigated the altercation.

And over the next twenty-four hours, Ian Trent had his personal investigators considered the incident. But to no surprise, their findings determined that Marcus was the guilty party. Though the disciplinary council members disagreed with the warden's findings, none dared cross Ian. The council of five sentenced Marcus to three days' forced labor, to be carried out at a nearby Krillium ore-processing center, thirty-five miles southwest of Runner's End.

#

Three days later, Marcus rested in his bunk back at Runner's End, exhausted from the grueling seventy-two hours of backbreaking labor he had endured at the ore-processing center. He also nursed bruised ribs and fists, injuries sustained while defending himself from the many attempts on his life while he was away. But Marcus had little trouble fending off the rank amateurs that seemed always to launch their attacks in groups of three or more.

A short time later, the pulsating energy bars of his cell deactivated. Marcus turned to see his cellmate enter, the very man he had saved four days earlier, Jackson. He was fresh out of the infirmary.

"As I live and breathe," said Jackson with a gruff voice and huge grin on his face, "if it ain't my guardian angel. What you did for me back there—"

"Was nothin'," replied Marcus as he climbed down from the upper bunk.

"C'mon now. Don't get all modest on me," Jackson said, limping toward Marcus. He extended his hand. "Name's Jackson."

At first, Marcus wanted to keep the silent loner routine going. But he changed his mind, figuring it wouldn't hurt to at least be affable, being that he was likely to spend the next eight years sharing a cell with the guy. He shook Jackson's hand.

"Marcus."

"Guess I owe you one, kid."

"Jackson, huh?" said Marcus, deciding to change the subject. "Is that a first name or last?"

"Just, Jackson. Nothin' more, nothin' less," said the purple-haired inmate as he took a seat on the lower bunk.

"So, what was that little scrap about?" Marcus asked.

"Just a misunderstanding. But forget all that. How can I make it up to you? You know, I'm sort of a big deal around these parts."

But Marcus dismissed the offer. The last thing he needed was some nutcase attached to his hip, feeling as if they were beholden to some crazy blood debt.

"Well, like it or not, you have a new friend in this place," Jackson continued despite his cellmate's objections. "Stick with me, kid. I'll show you what's what around here."

"Whatever you say, man," Marcus replied, desperate to end the conversation.

"So, how did you wind up in this little slice of paradise?" asked Jackson.

Marcus paused as he reflected on the actions of his so-called brother, Daren West, on Titan two years earlier. "I trusted the wrong person."

As Jackson inquired further, one of the human guards approached their cell.

"La'Dek, time for your session," said the guard as he deactivated the cell bars.

As tired as he was, the guard's words were music to his ears, as Marcus had no desire to dredge up old memories. He turned to his cellmate. "As much as I'd love to continue this discussion, I gotta go."

"Another time then," Jackson said as he watched the guard escort Marcus from the cell. He leaned back in his bunk while groaning from his injuries. He smiled despite the pain, grateful that there were still decent people like Marcus within the cruel confines of Runner's End.

#

Once again, Marcus found himself in the office of his caseworker, Charles Harris. As he began dozing off in his chair, which was cozier than the one in his last visit, Charles entered the room with his usual bundle of papers in hand.

"Sorry for the delay. You wouldn't believe the morning I've had," Charles said as he began organizing his files after taking a seat at his desk. He looked up at Marcus, noticing his face and hands. His countenance hardened with anger. "Who did this to you? Was it the guards? I want names."

"Don't worry about it, just gettin' to know the neighbors."

Charles could see right through the tough-guy act. However, he knew that it wasn't uncommon for incarcerated inmates to experience *turbulence* upon arriving to Runner's End, so he chose not to press the issue.

"Let's get started then," said Charles.

For thirty minutes, the counselor bombarded Marcus with a series of softball questions designed to get him talking.

During the conversation, they discussed everything from his motivation for first becoming a mercenary to where he saw himself following his incarceration.

It was clear Marcus didn't enjoy talking about himself. He spent most of the time dodging questions or giving vague and uninspired responses.

Eventually the ex-mercenary did open up, and in that moment, Charles switched gears, focusing on something a little closer to home.

"Tell me about your family, Marcus."

"I already told you about them."

"I'm not talking about your crew. I want to hear about your biological family," Charles clarified.

His real family was a topic he seldom discussed, even amongst his own crew. Feeling uncomfortable, he opted not to respond.

"It took some serious digging, but I see here you have a sister and a niece, Samantha and Elizabeth, right?" said Charles. "Care to talk about them?"

"There's nothin' to say."

"Do you stay in touch?"

"I used to."

"Why did you stop?"

"Let's just say, my sister is more than a little *frustrated* with me at the moment," said Marcus after a long, uncomfortable silence.

"Why do you think she's 'frustrated'?"

"You've seen my file, Doc. You tell me."

"Perhaps it's time you reconnect."

"Trust me, she wants nothin' to do with me," said Marcus, cutting him off. "Plus, our last conversation ended on less than a high note."

"How long ago was that?"

"'Bout four years back."

"Funny thing about time," said Charles as he leaned back in his chair. "It has a curious way of changing one's perspective... and even the *heart,* under the right circumstances."

"Or, it can make things worse," Marcus shot back.

"There's only one way to find out," said Charles with a smile. *And now comes the hard question,* Charles thought. "Do you miss them?"

Marcus paused again. He'd be lying if he said no. While irritated with Charles for even bringing them up, there was a huge part of Marcus that wanted to reconnect with his family. But it was far easier to simply shove them to the back of his mind, as he successfully had over the years.

For a long time, he used his profession to mask his pain. He hated knowing the decision he made for his life drove a wedge between him and his family.

But, of course, he had no desire to discuss these feelings with Charles, nor did he want to deal with the emotions that were beginning to surface.

"I've had a rough few days, you know?" Marcus said, refusing to make eye contact. "I need to hit the rack and get some sleep."

Charles could see that his simple question touched a nerve. He could tell that Marcus' family meant the world to him, even though he had a hard time expressing that fact. He didn't want to press his client too hard, so he backed off.

"You're right. We've covered enough for one day," Charles said as he stood to his feet. "We'll resume next week."

Marcus stood, relieved the session was over. He left the room toward the ASUs waiting for him in the hall.

As Marcus exited the room, Charles began to think. In his experience, people often needed a reason to forge a new path in life, and based on this conversation, he believed he may have just found that "reason" for Marcus.

He returned to his desk and activated his computer. It was time to track down Samantha. If he could somehow reunite the estranged siblings, he might just have a shot at breaking through to his client.

CHAPTER THIRTEEN

"Did I ever tell you how I ended up in this place?" Jackson asked from the lower bunk of the shared cell.

"No. But I'm certain you're gonna tell me," Marcus responded while resting atop his own bunk.

"You may find this hard to believe, but I'm not even supposed to be here."

"Yeah. You and everyone else in this joint," said Marcus, chuckling at how serious Jackson was.

"Well, my friend. Allow me to educate you on the stunning failure of the ISL's so-called *justice system*," said Jackson after sitting upright in his bunk.

As Jackson poured out his life's story, Marcus closed his eyes and listened. He figured that only about a quarter of Jackson's stories were true, but the tales were entertaining enough, so he often indulged his cellmate as he spouted his conspiracy-laced rants.

It turned out that Jackson wasn't the total nut job Marcus thought him to be. In fact, Jackson was a halfway decent guy, helping Marcus acclimate to his new surroundings for the past six months. Overall, Marcus was adjusting well given the circumstances. Though harassed by Razorback and his two lackeys, Sarak and Byrd, he had little trouble putting them in their place. Marcus even let Jackson join in the fun, allowing the scrawny inmate to get a little payback from their first altercation.

But there were other attacks; some so well organized, there was no way the likes of Razorback and his merry band of idiots could have arranged such assaults. They didn't happen often, but when they did, the altercations ended with Marcus taking the blame for the entire incident, while the perpetrators walked. During one attempt on his life, Marcus beat a partial confession out of one perpetrator. It wasn't much, but he learned that someone powerful orchestrated the attacks. However, the man confessing would rather die than speak the name of the true mastermind.

Marcus suspected the architect of the attacks to be Ian Trent, though proving that fact would be next to impossible. However, the guys thrown at him were way out of their league, allowing Marcus to continue with his day-to-day routine with little concern for his safety.

Other than the periodic pestering from Charles regarding the mounting injuries Marcus had been sustaining, the sessions were going well. For the first time, Marcus entertained the idea of a life outside the mercenary profession. Charles even talked him into reconnecting with his family by letter and the occasional voice transmission. While conversations with Samantha were often tense, he still looked forward to hearing from them, especially from his nine-year-old niece, Elizabeth.

As Jackson continued his epic yarn of tragedy and misfortune, Marcus examined another letter in his possession, it wasn't from his sister or niece, but from his dear friend and former crewmate, Skye. Though signed as anonymous, Marcus knew it was her. But he couldn't bring himself to respond. He needed to move on, and entertaining the thought of reconnecting with Skye, or anyone from his old crew, would undermine those efforts.

At that moment, a guard deactivated the cell bars.

"Time to report to your work details," said the armed guard.

"Guess we'll pick this up later," said Jackson as he got dressed.

Even though the guard arrived well ahead of the normal schedule, Marcus felt the guard's timing was impeccable.

"Just when the story was gettin' good," Marcus said as he too began preparing for the workday.

After they were dressed, the guard marched the two inmates to their assigned work detail in full view of the surveillance devices that Ian Trent was no doubt using to monitor Marcus' every movement. As he strode the grimy floors, Marcus discreetly extended his middle finger toward the cameras, sending a clear message to the warden as he walked through the exit of the housing unit.

#

Ian was furious at the display of sheer contempt. Every breath Marcus took was a personal affront to the warden. Ian took a deep breath to compose himself, then turned from the oversized view screen in his office to face his mistress and loyal prison administrator, Jazmin Ferrara. When it came to the operation of Runner's End, Jazmin was not only Ian's right hand, she was also the keeper and enforcer of his deepest and darkest secrets... secrets he couldn't utter in front of his best friend and campaign manager, Malak Pearce.

Jazmin was a cunning woman with an unnatural obsession for power. She hailed from Dellis City, the sprawling metropolis forty-nine hundred miles east of Cydonia, near the Gale Crater region of Mars. She had long, flowing, silver hair styled into a tight bun. On the left side of her neck was a serpent-like tattoo that seemed to move when observed from certain angles. While viewed as quiet and reserved to the uninitiated, her ruthless and strategic mind made her a rarity among the multitude of imprudent humans employed at Runner's End.

Her remarkable traits are what first attracted Ian to the mysterious woman, even at the cost of his own marriage. Ian was addicted to Jazmin, failing to see that her venomous words contributed to the emergent madness he'd developed over the years.

"It's been six months. Why is he still alive?" Ian asked.

"As I said, Marcus is a rare breed," Jazmin replied, staring at him with platinum-colored eyes with jet-black slit-pupils. "He has a warrior's heart. And it will take nothing less than a warrior to bring him down."

"Then how do we do this? As you know, 'warriors' aren't exactly an abundant resource in these parts," said Ian as he searched his antique wine rack for another bottle of his favorite Cabernet Sauvignon.

"Lucky for you, I had one transferred here from MPC-2312," said Jazmin. She passed Ian a small black holo-file.

Ian placed the file on his desk where it projected the holographic image of a former pirate and known drug runner. He had a battle-hardened look about him, with rough textured skin on the right side of his face, no doubt the result of a run-in with someone's white-hot Thermo-blade.

"And who is this gentleman?" asked Ian. Completely engrossed with the exquisite curves of Jazmin's body, Ian didn't even bother to read the accompanying dossier.

"Meet Jarek Malasen," said his mistress. "You're looking at perhaps one of the deadliest inmates ever housed in the Martian Penal Colonies in the last two centuries."

"Well, if he's that good, how the hell did he end up here?" Ian asked while pouring two cups of wine.

"I know how much you enjoy a good *bedtime story*," Jazmin said as she moved closer to her boss, caressing the side of his face with one hand while running her fingers through his dark hair with the other. "So, tell me how this one sounds..."

Jazmin explained that Jarek's capture happened two years earlier, during a raid on one of Teric Winters' narcotics facilities located in a secluded asteroid belt.

According to her sources inside the Interstellar Guard, Lieutenant Donald Shepard and his Shadow Tech Unit—Saber Team Seven—carried out the raid.

"Sounds interesting so far," said Ian, mesmerized by Jazmin's hypnotic, serpent-like eyes. "Please continue, Ms. Ferrara."

"It turns out that Jarek's youngest brother was among those killed during the raid. In fact, he died in Jarek's arms after Lieutenant Shepard fired the fatal shot," said Jazmin. She moved closer to whisper in the warden's ear. "Ever since that day, he's been obsessed with avenging his brother's death."

"Sounds like Jarek's quarrel is with the lieutenant. How does Marcus fit into this little equation?"

"Because the only reason Shepard found that hidden base was because Marcus tipped them off during his interrogations on Earth."

The look on Ian's face shifted from frustration to excitement, as this bit of unexpected news was a potential game changer.

"And while Jarek has no chance of getting his hands on Lieutenant Shepard, I'm certain he'll settle for the next best thing," said Ian as he pieced the elements of this fascinating story together in his mind.

"Exactly, my love. All you need to do is, *connect the dots*, as it were," said Jazmin as she took a sip of her wine.

"Where is Jarek now?"

"In the ultra-max security wing," Jazmin said, expecting the warden's question. "I've already prepared a squad of your elite ASUs to escort you at your leisure,"

"What would I do without you?" asked Ian, invigorated with a renewed sense of purpose.

He cut short his romantic rendezvous with Jazmin, as she knew he would.

"Perhaps it's time I met this Jarek Malasen," Ian said with a huge grin on his face.

"Your escort awaits," Jazmin said, motioning him toward his office door.

The time to strike is now, Ian thought as he exited his office toward the awaiting Android Security Units.

Once this bit of dark business was over, Ian could again shift his attention to the sole task that truly mattered, becoming the next president of the Interstellar League of Planets, so that he might fulfill the will of his liege, Teric Winters.

CHAPTER FOURTEEN

The ultra-max security wing of Runner's end was like something out of an inmate's worst nightmare. Enclosed in a climate-controlled bio-dome, the residents of this secluded area of the prison never saw the light of day. Hordes of RATs, the twelve-foot bipedal tanks, roamed the inner facility grounds with orders to kill any unauthorized personnel skulking about the prison. ARC units also canvassed the area, flying in tight formations while conducting ground-penetrating bio-scans, searching for those attempting to make a daring subterranean exit of the facility.

All potential escape routes were laced with cloaked smart mines that were capable of distinguishing inmates from prison staff. Once activated, the mines attached themselves to the target like a leech, injecting the victim with a liquefied version of white phosphorus, resulting in complete incineration from the inside out.

Inside the cafeteria at the center of the giant circular complex, Ian's ASU escort cleared the room of all inmates and staff. The android guards then moved all tables and chairs to the outer edges of the room, creating a large open space to prepare for Jarek Malasen's arrival. Ian stood with hands clasped behind his back while his elite ASUs positioned themselves in a semi-circle formation behind him. Moments later, the guest of honor arrived.

Jarek Malasen stood six-foot three with a muscular frame, including full sleeve tattoos on each arm. He had partially graying hair, shaved to near scalp level, and a stubbled beard also speckled with gray. His face carried the look of a man that had seen many battles, complete with the scars to prove that fact.

Bound hand and foot with indestructible shackles, Jarek walked into the room, escorted by two human guards armed with shock-batons.

Once face to face with Jarek, Ian dismissed the two escorts, having no desire for human ears to hear his proposal. For as much as he valued the need for sentient life forms staffing his prison, Ian didn't trust them very much. Therefore, to eliminate any chance of his words reaching the wrong ears, the warden would only speak to Jarek in front of his Android Security Units, whose memories could be selectively wiped following the conversation.

After the human guards left, the ASUs enclosed Jarek and the warden in a circle with weapons drawn in case the inmate made any unwanted movements.

"Mr. Malasen," said Ian as he moved closer to Jarek, "forgive me for not introducing myself upon your arrival. But I trust Ms. Ferrara explained—"

"Yeah, whatever," said Jarek. "What do you want?"

The warden would have ordered any other inmate beaten to within an inch of his life for such insolence. But Ian saw something different in Jarek's eyes. Behind the icy, murderous glare staring back at him, Ian could see the eyes of a true warrior without fear, as Jazmin first expressed back in his office. It was a look he'd seldom seen in all his years at Runner's End. In that moment, Ian knew that Jarek was the one for the job.

"I'll dispense with the pleasantries then," Ian said as he moved face to face with Jarek. "I've come with an offer you cannot refuse," Ian said as he began to circle Jarek at a slow pace. "I'm sure a man of your *considerable* talents has much to do outside these walls. And I assure you, your cooperation in this matter will hasten the day of your release."

Jarek had seen and heard it all during the twenty years he'd spent working as a pirate and mercenary for Orion's Shield. One thing his instincts taught him was to never trust a guy in a suit bearing gifts. He cast a wary glare at Ian.

"Ain't you that guy runnin' for president?" Jarek asked as it finally dawned on him where he'd seen Ian before. "Man, they said you were shady," he continued while looking around at the ASUs surrounding him. "Now I see why."

"*They* say a lot of things about me. But I perceive you to be a man that doesn't believe everything he hears."

"True. So how can I 'believe' you'll deliver on your end if I were to help with this little 'matter' of yours?"

"I've been called a great many things in my day, but a liar cannot be counted among them," said Ian. He reached into his inner coat pocket to retrieve a small circular holo-file containing a document recently submitted to him by Jazmin before the meeting.

"But to assuage your concerns, I had my assistant draft a copy of your release orders, signed by yours truly," Ian continued as he activated the device to project a holographic copy of Jarek's exit papers. "The only thing missing is the date. So, think about it, Mr. Malasen. Do you really have anything to lose?"

Jarek sighed as the orders displayed before him did little to change his initial opinion of the warden. But if there was one thing Ian had right, it was that Jarek had nothing to lose. If there were any chance of him getting out of that prison and back to hunting down Donald Shepard for the death of his brother, he figured it was at least worth hearing the warden out.

"Alright. Let's hear it," said Jarek.

"I knew you were a man of reason," Ian said with a sly grin on his face. "But in the interest of time, I'll cut to the chase. I need an inmate killed."

"Is that right?" said Jarek. "It's your prison, have one of your guards do it."

"The ISL government considers this inmate to be *high profile*. If I, as a presidential candidate, or any of my staff, were involved... well, I'm sure you know how that would end."

Jarek studied the look on Ian's face, seeing the eyes of a desperate man. But Jarek wasn't stupid, as this wasn't the first time someone made such a request of him.

In the past, those too rich and powerful to do their own dirty work had procured his services, lest they soiled their perfectly manicured hands. However, Jarek knew when he was being hustled.

His gut was telling him that if he were to take the warden up on his offer, he'd end up being a loose end.

"Forget it," said Jarek, laughing at the ridiculous proposition. "If I knock this guy off, I'll end up being made an example of, just so you can save face when the ISL comes knockin' at your door to investigate."

"Oh, but you have yet to hear the best part," Ian said, expecting Jarek's skepticism. "Let me tell you about your target, Marcus La'Dek."

Jarek's eyes widened with interest.

"La'Dek? He's here?"

"I take it you've heard of him?"

"There ain't many in our profession that hasn't," said Jarek, reflecting on Marcus' reputation. "But I also heard Teric has a bounty on his head so high, his own mother would give him up without a second thought. I don't know how a guy like that got under Teric's skin, but it must've been serious."

"If you'll indulge me. I'll tell you what happened. Perhaps you'll even reconsider our arrangement," said Ian.

"Go ahead. I'm sure this'll be good," said Jarek. He was dying to know what Marcus did to piss Teric off. He figured if nothing else came of this meeting, at least he would have his long-standing curiosity satisfied.

"First off, I know all about you, Mr. Malasen," Ian said. He began to circle Jarek once again in a slow, measured pace. "And I know what happened to your brother," he continued, watching the expression on Jarek's face transition from amusement to rage. In that moment, Ian knew he had Jarek's complete attention.

Ian recounted the events that led to the death of Jarek's brother, as explained to him by Jazmin prior to their meeting. As he spoke, he could almost see the anger growing within Jarek.

As he listened, all Jarek could see in his mind was the face of his brother, moments before taking his dying breath on that fateful day.

"And I'm sure you'd stop at nothing to exact vengeance on the man who murdered your brother," said Ian.

"I don't care how long it takes, Shepard and his family will die a slow, agonizing—"

"But what if I told you that Donald isn't the one you really want?" Ian asked.

Jarek lost it. He appeared to make a move toward Ian, but an ASU restrained him, though the android seemed barely able to hold him back.

"What are you talkin' about?" Jarek roared. "I saw him myself. He pulled the trigger."

"Come now, Jarek. You're smarter than that," said Ian. "Lieutenant Shepard is little more than a dog obeying orders. The real question is, how did they find that facility in the first place?" he asked. "After all, we're talking about a base that had eluded the all-seeing eye of the ISL for over fifty years."

"How the hell should I know?" Jarek yelled back.

"It was Marcus," said Ian.

Jarek was livid, yet he stood for a moment in stunned silence. Though they had little love for one another, the mercenaries of Orion's Shield lived by an unwritten code. One tenet of that code was to never sell each other out, especially to the ISL.

"Are you tellin' me that my brother is dead because La'Dek was a friggin' rat?"

"I'm afraid so," Ian said, as he braced himself for yet another eruption by Jarek. But to his surprise, he witnessed Jarek's face as it shifted from rage to an almost unnerving calmness. Then came the words Ian longed to hear.

"I'm in."

Jarek knew what he was getting himself into. In the end, he figured he'd be executed, or worse, thrown into stasis for the rest of his days. But he didn't care.

Regardless of the outcome, Jarek could go to his grave knowing the vow he made to his brother to kill the man responsible for his death would at long last stand fulfilled.

"Excellent," said Ian, as he once again moved face to face with Jarek. "Soon, I will transfer Marcus to ultra-max. I trust you'll find the appropriate time and place to address the situation."

Jarek nodded. Despite his anger, he kept his wits about him. He knew taking down the likes of Marcus La'Dek wouldn't be easy. Like a cunning predator, he would lie in wait until the opportune time to strike presented itself.

"Then we're done here," Ian said. He motioned for his guards to escort his guest back to his cell. As Jarek left the room, Ian reached out to Jazmin on his floating communication device. "He's onboard," Ian said. "Make the arrangements."

CHAPTER FIFTEEN

Jarek Malasen was above all else, a student of nature. He remembered long ago studying Earth's largest lizard, the Komodo dragon. Since the days of his youth, the tactic by which the dragon hunted had always fascinated him. The attack always began with an unassuming, yet venomous bite to the ankle of its prey. Over time, additional bites would come from other dragons. For weeks, the attackers would stalk their prey until their quarry collapsed, exhausted from both the constant harassment, and the debilitating venom coursing through its veins. It was then that the dragons moved in for the kill, with the Alpha receiving the largest share.

Jarek called this approach to hunting, *The Komodo Doctrine*. And it would be this very tactic he would use against Marcus to wear him down, setting up the opportunity for him to deliver the fatal blow.

Jarek's strategy yielded remarkable results. It had been three months since Marcus' transfer to ultra-max, and things were not going well. Without Jackson to watch his back, the constant stream of attacks were relentless. Everywhere he turned, random inmates looking to make a name for themselves took shots at Marcus. But to those bold enough to step up, he dealt stunning and embarrassing defeats.

On top of the increased attacks, the warden cut Marcus off from his weekly sessions with Charles, whom he relied upon to help keep a level head. The constant harassment and lack of support proved to be a toxic combination. The weight of his situation began to take a physical and emotional toll on his body and mind. For months, Jarek orchestrated the attacks from the shadows until, finally, the sinister mercenary made his presence known.

As Marcus sat by himself at the beige-colored table in the cafeteria, Jarek approached. He sat on the bench opposite Marcus. For minutes, Jarek stared at his target. But Marcus never looked up as he continued his dinner. Eventually, Marcus raised his head, barely recognizing Jarek's face. Since his transfer to ultra-max, the two never spoke, despite being neighbors on the same cellblock.

"Look, man, I'm *not* in the mood," said Marcus. "So, do yourself a favor and keep it movin'."

"I see you're not feelin' talkative today, so I'll keep it short," said Jarek with a piercing glare fixed upon his prey. "You did *somethin'* that got *someone* very close to me killed."

"Yeah? How do you figure that?"

"You broke the code."

"What are you talkin' about?"

"Two years ago, while you were on Earth, singing your little heart out to the ISL, you gave up the location of Harbinger Base."

"So what? Teric pulled the plug on that place years ago. It was nothin' more than a glorified warehouse, storing product he couldn't even move."

"That would be a negatory," said Jarek. "Teric had just reactivated that base. Hell, we were prepping for a major score when Shadow Techs raided the place."

Marcus knew there would be collateral damage when he cooperated with the ISL. He just never expected to be staring at one of those casualties so soon.

"I didn't know..."

Jarek banged his fist on the table, nearly breaking it in two.

"You had no business runnin' your mouth in the first place," said Jarek. "And now, *my brother* is dead, because of you."

The sum of everything that Marcus had been going through was bubbling to the surface. And before he realized it, he snapped, only exasperating the situation.

"Look, I'm sorry for your loss. But unless you wanna join your brother tonight, you'd best step away from this table," said Marcus, returning the icy glare to Jarek.

Enraged, Jarek stood to his feet and grabbed the edge of the table, nearly flipping it on top of Marcus.

"You arrogant bastard," roared Jarek.

The entire room fell silent. All eyes were on them.

"I'm right here," said Marcus as he moved face to face with his opponent.

After a momentary stare down, Jarek regained his composure.

"All deeds come with a price, La'Dek," said Jarek. He began to walk away, then turned toward Marcus one last time. "I'm Jarek Malasen, by the way," he said. "Just wanted to introduce myself. You know, before I killed you."

"You know where to find me," said Marcus as he watched Jarek walk away.

As Jarek departed, three members of his old crew, Soren, Alec and Jalen, approached. However, Marcus went on the offensive, not even waiting for Jarek's crew to attack. Marcus approached Soren and delivered a ruinous low kick, shattering the seven-foot-tall inmate's knee. He followed up with a ridge hand strike to the throat. The towering man fell to the ground, clutching his knee, yet unable to cry out due to his fractured larynx. Marcus started to finish him but opted rather to let him squirm on the ground with no one to come to his aid.

Jalen grabbed Marcus from behind, allowing Alec to land a series of left and right hooks to his head. But Marcus delivered an enervating kick to the groin of Alec, causing him to cease the assault. Though still locked in Jalen's crushing bear hug from the rear, Marcus used a tried-and-true jujitsu technique to break the hold. Marcus thrust both of his arms in an upward motion, disrupting Jalen's grip. It wasn't enough to break the hold but created enough space for Marcus to shift his hips and legs to the left, while leaning forward.

Marcus positioned his right leg behind Jalen's. Using his right arm, he drove Jalen's hips back while at the same time thrusting his right leg forward. The move caused Marcus to hit the ground first, taking Jalen along with him, breaking the hold altogether. Marcus then rolled into a full mount position over Jalen, where he beat the man to within an inch of his life.

Marcus could hear Alec charging from behind. He rolled off Jalen and caught Alec with a leg sweep, sending the muscle-bound mercenary crashing to the floor. Marcus stood over Alec and began stomping him into the ground. By now, everyone in the cafeteria surrounded the combatants, cheering them on. Even the human guards watching the prisoners stood by, allowing the carnage to continue.

Just before Marcus could stomp Alec to death, Jarek joined the fight, delivering a heavy blow to the back of Marcus' head that sent him reeling to the floor. Despite nearly blacking out from the punch, Marcus staggered to his feet, as if operating on sheer instinct alone. As Jarek rushed in to deliver another haymaker, he stopped in his tracks upon hearing the booming sound of a weapon fired into the air.

"Stop," cried a voice very familiar to Marcus.

As Marcus struggled to shake off the effects of the punch, he looked to his left. To his surprise, he saw his counselor, Charles Harris, pistol in hand, standing next to the Overseer of the ultra-max security wing, Commander Gordon Hastings.

For the past three months, Ian and his administrative staff had stonewalled Charles, keeping him from seeing Marcus. Every time he demanded to conduct a welfare check on his client, they met Charles with nothing but excuse after excuse, each more absurd than the last.

After threatening to invoke the help of several powerful ISL planetary senators, with whom Charles was a close friend, Ian reluctantly allowed Charles into the ultra-max security wing to check on Marcus.

When he finally arrived at Commander Hastings' office, Charles saw the horrific scene taking place between Marcus and Jarek's crew on the nearby view screen. The entire sight resembled some sort of primitive gladiatorial combat. Appalled that Commander Hastings would allow such a travesty to happen under his watch, given his distinguished career and reputation, Charles threatened to have the Overseer fired and brought up on charges if he didn't take him in there to break up that fight.

Upon reaching the cafeteria, Commander Hastings refused to move with a sense of urgency, that's when Charles took matters into his own hands. So, he snatched the pistol from Commander Hastings' holster and fired it into the air, bringing an abrupt end to the fight just before Marcus and Jarek killed each other.

Charles returned the pistol to the supervisor.

"Commander Hastings. Get these inmates out of here," said Charles.

"Don't tell me how to run my—"

"Now, Commander," Charles demanded. "Or I promise, I'll have the ISL in here by the end of the week to shut you down."

The commander complied, knowing full well of Charles' extensive contacts on Earth and abroad. As the commander ordered his staff to clear the room, Marcus collapsed to the ground as the exhaustion from all that he'd endured over the last three months had caught up with him. Charles moved in to help Marcus back to his feet.

"Let me guess, you happened to be in the neighborhood?" said Marcus, trying to catch his breath.

"You know me," said Charles. "I try to get out when I can. Are you alright?"

"I'm fine," said Marcus while glaring at Jarek. The two men locked eyes as the guards escorted Jarek from the area.

Charles could see that look in Marcus' eyes. "I'm having you transferred to the infirmary, under the watch of guards I trust. And it's time we resumed our sessions too."

"For what?" Marcus yelled in frustration. "There's only one way this thing ends."

Charles sighed as he motioned for the medics who had just entered the cafeteria. As the medical personnel strapped Marcus to a levitating gurney and moved him toward the infirmary, Charles turned to Commander Hastings, shaking his head in disappointment.

"What happened to you, Gordon?"

"I fell in line, Charles, which is more than I can say for you."

"You're a disgrace," said Charles. He turned his back to the commander and headed toward the exit.

"Why don't you go back to Earth with the rest of those self-righteous Haven City bureaucrats?" said Gordon.

But Charles ignored the commander and continued toward the exit. At that moment, he had bigger concerns. It seemed as if all the progress Marcus had made since his arrival to Runner's End was coming undone. Charles knew that if he didn't act, Marcus may plummet down a hole from which the embattled former mercenary may never return.

CHAPTER SIXTEEN

"I've had it with this place," Marcus yelled to Charles.

After two days in the infirmary, the guards brought Marcus to his counselor's office. And for the last thirty minutes, Charles had allowed him to vent. It was a necessary move to keep his client from exploding under the immense pressure that had been building since his transfer to ultra-max.

"I don't care what you say, Jarek is dead. And when I'm done with him, Ian Trent is next."

"C'mon, Marcus, that's what they want. You're playing right into their hands"

"I don't care," Marcus roared. He moved about the room, looking for hidden surveillance devices. "I know you're listenin', Trent. You're next. You hear me?"

"There's no listening devices here," said Charles. He moved toward Marcus to keep him from ransacking his office. "I swept the place myself. What you say here stays here."

"It all makes sense now. I was never meant to leave this place alive. And now, I don't care what happens."

"Don't you dare say that," said Charles, "I've had a lot of folks walk through that door over the years. Many of them were true lost causes. But not you. I've seen your ability to—"

"You know," said Marcus, "Jarek mentioned something back there. He said, *All deeds come with a price.*"

Charles started to interject, but Marcus continued.

"Maybe this is my penance for all the dirt I did over the years," added Marcus as he turned his back to Charles. "You say those guys out there are lost causes? Well, last I checked, I have just as much blood on my hands as—"

"We don't always have to *be* what we've *been*," Charles interrupted. "Right now, you have a choice."

"What *choice* do I have? Wanna know my choices, Doc? Here they are, kill or die, that's it."

"C'mon, Marcus."

"I'm done talkin'. I'm tired of holdin' my peace. Tired of lookin' the other way. They wanna see the real me? Fine, they can strap in and watch as I burn them and this prison to the ground."

Charles could see there was no talking him down as he watched in dismay as Marcus spiraled to a dark place. It was time to bring in the reinforcements. Marcus continued his expletive-laced description of the atrocities he planned to visit upon Ian and Jarek. As he spoke, a familiar voice came from behind. The voice brought a sudden halt to his gut-wrenching tirade.

"Marcus," said the strong, yet gentle voice.

Marcus turned to see his sister—Samantha—standing in the doorway. She entered the room from the seldom-used waiting room attached to the counseling office. Charles stood silently by her side.

Her presence stunned Marcus. Thinking she was some sort of hallucination, he stood in silence.

Samantha was ten years older than Marcus, though she looked much younger than her forty years of age showed. She was a strikingly beautiful woman with a toned, athletic build. She was of a darker complexion than her brother, with black, flowing hair tied back into a simple ponytail.

Marcus didn't know how Charles pulled this off, but in that moment, it didn't matter.

"Sam," he said.

To Samantha, he almost sounded as he did when he was a child.

"What are you doin' here?" he continued.

"Making a long overdue visit to my baby brother," said Samantha with a radiant smile.

For a moment, Marcus forgot all about Jarek, Ian, and the rest of his troubles. But he was ashamed to let his sister see him like this. "You shouldn't be here right now."

But Samantha ignored her brother and instead embraced him.

As she held him tight, something happened that the battle-weary mercenary never experienced in his entire life, he broke down.

Hearing her brother wail in pain brought tears to Samantha's eyes. She looked back to see that Charles had excused himself to the waiting room, allowing them to talk in peace.

"How did we let it get like this?" asked Samantha.

"I don't know," said Marcus. "Look, Sam... I... I'm sorry..."

"No, Marcus," Samantha said, not even letting him apologize. "*I'm* the one that's sorry. I shouldn't have given up on you," she said while wiping the tears from her brother's eyes. "Now that Dad is gone, we're all we have. And I'm not goin' another day allowing this family to remain broken."

Samantha's words brought life back to Marcus, allowing him to at long last breathe again.

"Where's Liz?" he asked.

Samantha paused before answering. While she didn't want to bring him down, Marcus needed to know the truth.

"She's in the hospital, on Stratus One," said Samantha. For the next ten minutes, his sister brought Marcus up to speed on all that had been going on while he was away from the Stratus One Star Port, their childhood home on the edge of ISL-controlled space.

Samantha told him of the unknown disease with which his beloved niece had been suffering. But she also expressed the love that Elizabeth had for him. Marcus' heart broke as he listened to what Elizabeth had been enduring since birth.

"She gave me a message to give to you," Samantha continued. "She said, 'It's time for you to come home.'"

Marcus felt ashamed. If his niece could find the strength to fight, how could he have the audacity to give up? He couldn't articulate the things he wanted to say, so he nodded his head in agreement.

"People like this Jarek and Ian that Charles told me about? They'll have their day. God will not let injustice stand, you know that," she said, reminding him of lessons he learned long ago. "You just need to concentrate on coming home to us. We need you."

"I promise," said Marcus amid a flood of emotions. "I'm comin' home. You have my word."

"That's all I need to hear," said Samantha. She put her gloved hand on the side of his face. "We'll be waitin' for you, no matter how long it takes," she said. "Stay safe out there, baby brother."

"I always do," replied Marcus with a rare smile on his face.

She gave her brother a kiss on the cheek then met up with Charles in the waiting room.

As Marcus stood in silence, he felt a sense of peace. He resolved within himself to keep the promise he made to his sister and niece. No matter what it took, he would not let Jarek and Ian win.

#

For the next three weeks, Marcus did everything he could to avoid Jarek. Amid the taunts and other provocations, Marcus did an amazing job of keeping himself from killing his adversary. Charles had increased their sessions to three times per week, despite the warden's protests. Things seemed to be getting back to normal. It wasn't until one night the very next week that everything went sideways.

As Marcus entered the exercise area for his evening run, he noticed that everything seemed wrong. The entire room was empty. But the facility was not just empty of inmates; there were no guards present, human or android. His instincts told him to get out of there. But as soon as he turned to leave, the automated blast doors, which were never closed, closed shut with a resounding thud. He could hear the locking mechanism engaging.

Ian had set the entire moment in motion, as he grew tired of waiting. He wanted Marcus dead, so he urged his assassin to move up the timetable. Moments later, the staff door opened on the far side of the room.

Marcus half expected to see the likes of Ian or Commander Hastings emerge, spouting some pre-rehearsed speech, flanked by a few bodyguards. But to his surprise, it was Jarek Malasen only. The menacing mercenary approached slowly with a thermo-blade in hand, glowing red-hot, with the look of a man obsessed. Still, Marcus tried to talk him down.

"C'mon, Jarek. We don't have to do this," said Marcus. He didn't want to kill his foe, knowing that if he did, Ian would have him executed for murder.

"That's where you're wrong, La'Dek," said Jarek. "You see, I'm a man of my word."

"What did Ian promise you? Freedom?" asked Marcus. "He's playin' you, Jarek," he continued. "You've been in the game long enough to know how this ends."

"Yeah, I know exactly how this ends," said Jarek as he twirled the knife between his fingers. "This ain't got nothin' to do with that little worm, Ian," he said. "*This...* is for my brother."

Before Marcus could say another word, Jarek went on the attack, unleashing a series of brutal slashes from random angles.

With quick reflexes, Marcus dodged the onslaught. As he bobbed and weaved out of harm's way, he could feel the searing heat from the knife as it passed mere fractions of an inch from his body.

Jarek's knife techniques were flawless. He made short, compact slashing and piercing strikes, while switching between standard and reverse grips.

It had been quite some time since Marcus faced an opponent as skilled as Jarek, but he was up for the challenge.

While Marcus could dodge and thwart many of the attacks that followed, some were getting through to his forearms and shoulders. Though most of the strikes that landed grazed him, the pain was sharp because of the searing heat of the blade.

Each time Marcus came close to disarming him, Jarek switched hands with the knife just before Marcus could lock in a hold.

While Jarek would've carved any other person to pieces with his assault, Marcus survived the initial attack with only minor cuts and burns.

Marcus could see Jarek's confidence level rising with each attack. But Marcus' experience taught him to maintain discipline and wait for his opponent to slip up. And that is just what he did.

Jarek's overconfidence got the better of him. He attacked with an overhead slash.

Marcus stepped toward Jarek while grabbing his arm midair.

With full control of Jarek's arm, Marcus stepped in for an over-the-shoulder throw that planted Jarek flat on his back, allowing Marcus to pry the knife away.

Marcus wanted to use the knife to finish Jarek off but thought better of it. He couldn't afford any mistakes.

"I don't need a weapon," said Marcus as he stepped back and tossed the thermo-blade across the room.

"Same here," said Jarek as he arose and rushed toward Marcus, swinging with a series of lefts and rights.

Marcus parried most of the incoming blows with his arms, then stepped in, grabbing Jarek by the throat with both hands. With a sudden burst of adrenaline, he lifted Jarek from the ground and pinned him against a nearby wall.

Jarek could feel himself starting to black out. Using brute force, he pried Marcus' hands away from his throat just enough to slip out of the hold. He stumbled to his knees, gasping for air.

Marcus seized the advantage by delivering a devastating kick to his opponent's torso. The blow lifted Jarek a few inches from the ground, fracturing his ribcage. Marcus began wailing on Jarek with a barrage of hooks, uppercuts and elbow strikes, prompting Jarek to cover up to keep Marcus from knocking him cold.

Despite the pain from his injured ribs, Jarek fought back. He blocked one of Marcus' incoming haymakers, allowing him to transition into a hip throw of his own, slamming his adversary to the floor. Jarek mounted Marcus and began launching his own assault. But Marcus covered his face with a forearm guard to minimize the incoming damage.

Marcus grabbed the sides of Jarek's shirt. Using every ounce of remaining strength, he shifted his attacker's torso to the left, allowing him to roll Jarek off him. The two men scrambled back to their feet. Marcus tried in vain to convince Jarek to stop.

"It's time to end this," said Jarek. "Only one of us is walkin' out of here alive."

Marcus could see in Jarek's eyes that he would fight to the death. He wanted to keep the promise he made to his sister, but his anger began to get the best of him. All the torment that he experienced for the last three months flooded his mind.

"You're right," said Marcus. "One of us *will* die tonight. But it ain't gonna be me."

Enraged, the two men charged one another. They met in the center of the room, trading hellacious blows well into the night. Any lesser man would have crumbled from the dreadful shots landed. But each man remained on their feet, as if by force of will alone.

As the fight went on, the sum of everything that Marcus endured throughout his life filled his mind, only fueling his rage all the more. He pressed the attack. He could see that he was wearing Jarek down.

Jarek's arms gave out from extreme exhaustion. But Marcus pressed the assault, knocking his opponent onto his back. He rolled Jarek onto his stomach and locked in a rear chokehold, while locking his legs around his opponent's waist. Jarek tried to break the hold, but Marcus locked it in even more. There was no escape.

"Do it," grunted Jarek. He was ready to join his brother, so he stopped struggling.

As Marcus applied pressure around Jarek's throat with his bicep, he could almost see the man's life draining from his body. Every fiber in his being wanted to snap Jarek's neck in two, but he stayed his hand. He began to think about Samantha and Elizabeth, and how they needed him back home. More than that, *he* needed them. Marcus released the chokehold and kicked Jarek away. He staggered to his feet and stood over his fallen foe. Jarek was now on his hands and knees, gasping for air.

"You should've finished me," Jarek said amid violent outbursts of coughing and wheezing. "I'm never gonna stop comin' for you."

"I know," said Marcus as he looked down on the dejected mercenary on his hands and knees.

With a sudden and powerful downward elbow strike, Marcus crushed Jarek's lower vertebra, paralyzing him from the waist down.

Jarek cried out in agony.

"Now, it's over," said Marcus.

Marcus looked up at the surveillance device in the room in contempt, knowing full well that Ian was watching. He put his hands behind is head and dropped to his knees, waiting for the guards to enter. Moments later, Commander Hastings and a squad of Android Security Units stormed the room, with rifles fixed on Marcus.

#

Ian was furious as he watched his best-laid plan come to a sudden and violent end on the eighty-inch view screen in his office. Sure, the medical technology to repair Jarek's obvious spinal injury existed. But he'd have to have the man transported to the Luna One hospital on Earth's moon to do so.

It would be months, if not years before Jarek would recover from his injuries. Even then, he wouldn't be the same. As much as Ian hated it, Marcus was right, *it was over.*

Ian turned to Jazmin.

"Perhaps it's time for Plan B," said Ian.

"And what, may I ask, is 'Plan B'?" said Jazmin.

Ian turned his attention back to the view screen and watched as his officers took Marcus into custody. With a sly and sadistic grin on his face, he turned back to Jazmin.

"What else, my dear?" said Ian. "*Noctis Labyrinthus.*"

Jazmin smiled. "*Very* good," she said. "I'm on it."

CHAPTER SEVENTEEN

Ian Trent sat behind the rare, exceedingly expensive oak wood desk in the corner of his office, with furrowed brow and clenched fists. The warden listened to the Chief Physician of Runner's End as she delivered to him the rather extensive list of injuries sustained by his vile assassin, Jarek Malasen.

"He should've saved me the trouble and just finished him," muttered Ian under his breath. Ian signed the off-world transfer documents and sighed.

"Excuse me, sir?" asked the Physician in Chief.

"Nothin'," said Ian as he shoved the paper-thin data pads back into her hands. "When does he leave?"

"As soon as we're finished here," she said. "I know Mr. Malasen's spinal injuries are severe, but I assure you, sir, the Luna One hospital has some of the brightest minds in the—"

"Yeah, that's great," Ian interrupted. "Let's get this underway, shall we?"

The doctor paused, offended by the warden's brash tone. She started to speak, but Ian once again interrupted.

"And, Doctor, I trust you'll exercise the utmost *discretion* in this matter, yes?"

The doctor nodded her head uncomfortably. In that moment, all she wanted was to get out of that office. After glancing over the signed documents, she quickly turned and departed the room. As the automated door slid shut behind her, Ian began tossing furniture across the room whilst shouting obscenities into the air.

The weight of the task given to him by his liege, Teric Winters, was feeling almost too much to bear. And Marcus' decision to spare Jarek's life only added to the ever-growing list of problems plaguing his every waking moment. Ian stared blankly across the room, furious that he found himself with the unappealing task of having to tie up yet another loose end.

Sure, he could have spared Jarek that final spine-crushing blow that ended the savage brawl. But in Ian's mind, there had to be a reckoning for his assassin's staggering failure. So, he allowed the fight to crescendo to its climactic yet catastrophic end by delaying the call to send in his guards to break up the fight. Ian was certain that Marcus would kill Jarek, which would have made his life much easier, but he was sadly mistaken.

Ian reached toward the antique wine rack next to his desk and poured himself a tall glass of wine from the first bottle he touched. He moved the expensive goblet to his lips and began guzzling the drink as if it were a glass of water. As Ian tried desperately to drown his sorrows, he began feeling the pulsating headache that had been building all day.

The throbbing pain grew only worse as he began to think upon the enormous pressure he'd soon face by Charles Harris and his Earth-based associates in the aftermath of what they would later call *The Malasen Incident*.

He had earlier sent Jazmin Ferrara, his most trusted advisor, to enact what he called, for lack of a better term, *Plan B*. As he awaited her call to inform him that the preliminary details of that plan were in place, he returned to his chair to ponder his next move.

While he had yet to work out the details of his latest scheme, he knew the exact location where it was all to take place, the most treacherous labor detail on all of Mars, the dreaded work camps of *Noctis Labyrinthus*.

#

"Alright, son," said Charles amid a sigh of frustration. "Let's get this over with."

Charles and Marcus exited the counselor's office, with two Android Security Units following closely behind.

"You ready?"

"Not like I have a choice," responded Marcus, motioning toward the uncomfortable shackles locked around his wrists.

"I need you to keep it together in there today," said Charles.

They started the slow walk down the dimly lit corridor toward the disciplinary board's chamber in the administrative wing of Runner's End.

"Relax, Doc," said Marcus. "This ain't my first go around, you know."

Charles could almost hear the surrendered tone in Marcus' voice. He longed to do more, but it seemed they were fighting a losing battle from the beginning. He only wished he'd seen it sooner.

As they strode the stone floors of the hallway, they headed toward the last of a month-long series of hearings held before the prison's disciplinary board. Supposedly an independent body, with no personal ties to the warden nor any of his staff, the government tasked the board with investigating all incidents at the prison, allowing them to deliver impartial punishments to those involved if necessary.

However, it seemed clear from the very first hearing that the board members were little more than Ian Trent's echo chamber, spouting the same nonsense that Ian had been spreading for weeks.

Throughout the hearings, Ian argued that Marcus deserved harsh punishment for the unprovoked attack on his fellow inmate, Jarek Malasen. Charles thought it absurd how the warden painted Malasen to be some helpless, upstanding prisoner that was a victim of sheer circumstance. Even worse, the degree to which the five board members fell for the steaming load of excrement the warden had been shoveling their way absolutely stunned Charles.

Unbeknownst to anyone, it also took the careful *persuasion* of Jazmin Ferrara to sway the collective decisions of the board members. Jazmin's behind-the-scenes endeavor paid off. And perhaps out of fear for the lives of their families, the council agreed by a unanimous decision, sentencing Marcus to four standard Earth years' hard labor at the work camps of Noctis Labyrinthus.

Jazmin's dark methods were so flawless that even Charles could not legally overturn the board's decision. In fact, the only concession that Charles squeezed out of the warden was to allow him and Marcus to continue their counseling sessions remotely. It wasn't much, but it was the best Charles could do given the circumstances, and it was the only way he'd be able to check on the wellbeing of his friend and client.

Within ten minutes, Marcus and Charles found themselves inside the cold confines of the small, unimpressive boardroom. The centerpiece of the room was a long *synthwood* table made from a cheap synthetic wood, more durable than its genuine counterpart. From plush chairs situated along one side of the drab, brown-colored table, the board members stared at Marcus with stern, judgmental eyes as they prepared to begin with the morning agenda.

The hearing itself was little more than a formality, as the board had already informed Charles and Marcus of their decision hours earlier. As Marcus sat on the lone metal bench in the room while they stated his sentence for the record, Charles sat by his side, shaking his head in dismay.

"Don't worry, Marcus. We'll fight this. I'll have my people—"

"Forget it," said Marcus. "Trust me, I'm more than happy to leave this place."

"You don't understand," said Charles. "The conditions at the Noctis Work Camps aren't fit for a—"

"I said don't worry about it," repeated Marcus as he stood to his feet. Moments later the guards approached, once again shackling his arms and legs. "I can handle myself."

As the guards escorted him toward the exit, Marcus turned one last time toward his counselor and friend before departing the room. "See ya around, Charles."

CHAPTER EIGHTEEN

Though he once again found himself en route to a strange and unforgiving environment, Marcus was relieved to escape the unending drama he endured inside the main prison. Unlike the rowdy prisoners aboard the transport that first brought him to Runner's End a year earlier, the crowd on the transport headed to the Noctis work camps was eerily silent. To Marcus, the vibe coming from his fellow passengers was like inmates being marched to their death, awaiting their turn inside bloodcurdling execution chambers.

Even the skies seemed to darken as they approached that terrifying place. Powerful thunderclouds swelled over the region, complete with high winds stirring up dense, rust-colored dust storms on a massive scale. Despite the tempest raging around them, Marcus was sound asleep inside the armored hover transport, taking full advantage of the eight-hour trip to the unknown perils awaiting them in the distance.

Hours later, the blaring sound of the navigational alert system awakened him, signifying their transport was approaching its destination. After coming to a complete stop, the guards marched the prisoners from the rear of the transport in dual columns of eight. Once outside, the inmates stood in silence as they waited for their new overseers to arrive and process them into the camp.

Marcus looked about his surroundings only to see an unforgiving, windswept wasteland, seemingly devoid of all life. The area was a stark contrast to the rest of Mars, which had been restored to life through a millennium of terraforming.

The only visible constructs in the area were the large, metallic prefab structures, rectangular in appearance. The makeshift buildings made up the entire mining camp and provided only the most minimal protections from the turbulent weather that frequented the area.

"Welcome to Work Camp Noctis," said one of the human guards who marched the inmates from the rear of the vehicle.

The guard spoke from behind a jet-black respiratory mask, giving his words a filtered yet amplified tone, allowing his voice to compete with the blistering winds howling throughout the area. He, like the four guards accompanying him, carried pulse-rifles and shock batons. And they had no compunctions about using them.

From the inmate's perspective, the faces of the guards seemed almost as featureless as the Android Security Units swarming throughout the camp. The combination of the guards' facemasks, advanced image processing goggles, and head-to-toe metallic body armor covered all traces of their humanity.

"The Overseers will be here soon," continued the guard in his almost robotic tone. "You will not speak unless spoken to," he said while pacing the ground before the prisoners. "Step out of line, and I promise you this, you *will* die out here."

Marcus was relieved he didn't have to sit through another long-winded, Ian Trent-inspired speech intended to frighten the already petrified inmates. One look at their surroundings and everyone knew, there was nowhere to run. As they awaited the Overseers, Marcus turned to his left to see he and the others were standing atop a ridge overlooking a vast network of canyons below, from which the region no doubt received its name.

For *Noctis Labyrinthus* was a widespread, maze-like system of canyons near the Martian equator, better known as *The Labyrinth of the Night*. Situated along the western edge of the 2,500-mile-long Valles Marineris Canyon, the deep features of both Noctis and Valles were so pronounced upon the planet's surface, one could clearly see the entire region from space, like an ancient scar stretched across the heart of Mars.

Tasked with extracting newly discovered pockets of Krillium ore from the region's intersecting valley walls, Marcus and his fellow inmates were sent to the area to link up with a larger mining force comprised of inmates and their heavily armed Overseers. It was dangerous work to be sure, as frequent landslides plagued the valley to where even their localized shielding system struggled to protect the workers below.

As Marcus continued to survey his surroundings, he noticed that security wasn't as tight as it was at Runner's End. He was later told that the reason for the reduced guard presence was because there was little to no vegetation, edible or otherwise, in the area. To make matters worse, there was no water in the entire region. All provisions were flown in from thousands of miles away and locked in an impenetrable vault, rationed out to the miners daily by their taskmasters.

The very notion of running was a true death sentence. And those bold enough to venture into the maze of canyons below were never heard from again. There were even rumors of ancient beasts lurking throughout the dark recesses of the canyons, which inmates could almost hear in the dead of night. The entire land was a forbidding place, and as the first of their Overseers arrived, the area became even more uninviting.

Comprised of Android Security Units and the native Martian species known as the Ekron, the force called *The Overseers*, were notorious for their brutal tactics and heavy-handed leadership style. While the ASUs ran the overall mining operation, the slender, yet exceedingly strong, Ekron served as their merciless enforcers.

Adorned in spiked armor, reminiscent of medieval horrors described in ancient Earth mythology, the Ekron had oversized, solid black eyes and coarse, gray skin, tougher than any leather known to man. The Ekron's menacing appearance and ferocious nature was often enough to keep the inmates in line, and their presence seemed to have the desired effect on the newly arrived prisoners. As the Ekrons marched the inmates toward the processing facility in the distance, Marcus began to wonder if he could even keep the promise he made to his sister and niece.

He needed them and longed to return to his childhood home on the Stratus One Star Port to be with them. But four years was a long time. He figured it was just a matter of time before Ian sent another assassin his way, or worse, he became the latest victim of the environment itself. After all, there were no shortages of ways to die, in the *Labyrinth of the Night*.

CHAPTER NINETEEN

"Alright, guys," yelled Marcus from the rocky base of the Noctis Canyon floor to the three inmates standing on the ridge above. "Bring it down."

At the prior command of the Overseers, they rigged a three-ton laser drill with cables attached to a heavy-lift droid (HLD), with orders to lower the massive piece of mining equipment to the ground below. Resembling some headless animal, the HLD walked on all fours, with a high tensile-strength cable-support system attached to its broad back.

One of the droids' operators started the stabilization protocol. The mechanical beast anchored itself to the surface by driving massive spikes attached to its sturdy legs deep into the ground. They maneuvered the drill over the ledge and began the time-consuming task of lowering the weighty piece of equipment to Marcus and his crew below. The cables creaked and the droid shook as they fought dangerous winds whooshing throughout the canyon, winds that seemed determined to keep them from accomplishing the hazardous duties assigned to them by their thoughtless taskmasters.

Irritated because they had to lower the drill in this manner, Marcus wished there was another way. But the only flight-capable vehicles available to them were the small ARCs, typically used to detain prisoners. However, the ARC's small size and weak propulsion system proved a disastrous combination when, a week earlier, strong wind gusts sent two of them crashing into the valley walls, destroying the expensive laser drill they were hauling.

The larger personnel transports, built with thrusters capable of operating in high-wind environments, were all grounded for maintenance.

So, instead of waiting, the ASUs ordered the manual lowering of all equipment to the valley floor, despite the considerable risks to the wellbeing of the inmates.

"And why are we usin' these outdated drills, anyway?" asked an inmate named Laris as he and Marcus watched the drill make its unsteady descent into the canyon. "Ain't this what mining drones are for?"

"Yeah, but then it wouldn't be much of a *forced labor camp* if they used those, right?" countered Marcus.

The two men held their collective breath as they watched the drill sway in the wind while the inmates above adjusted the heavy-lift droid's rate of descent. They were losing daylight, and everyone knew they'd face severe consequences if they dropped the drill, so they worked in unison to avert disaster.

According to the Galactic Time Unit (GTU) calendar, a standard measure of time to which all member races of the Galactic Planetary Consortium adhered, it had been four full cycles since Marcus' arrival to the work camps of Noctis Labyrinthus. By Earth Standard Time, four full cycles were just over four and a half years, an eternity in Marcus' mind, and over six months longer than his original sentence.

Marcus noticed a disturbing pattern during his extended stay at Noctis. The inmates never had suitable equipment to carry out their unrealistic duties. And the Overseers seemed more concerned with filling quotas than preserving the lives of those tasked with fulfilling those orders. But Marcus pushed those thoughts from his mind. He had only two months left at Noctis and didn't want to partake in any activity that might prolong his stay. Much to his surprise, Ian never sent another assassin his way. In fact, all attempts on his life had ceased, though the perilous environment did try to claim his life on countless occasions.

Though he had no desire to even breathe the same air as Ian Trent, there was a small part of him that looked forward to his return to Runner's End, if for no other reason than to once again sleep in a proper bunk.

He snapped out of the daydream of better conditions and continued to watch the drill as it inched ever closer to the valley floor. Twenty minutes later, they finished the job without incident. Marcus and his four-man crew maneuvered the drill using its hover-lift thrusters to the target position, under the watchful gaze of two ASUs and three Ekron enforcers.

"Change of plans," said one android. "We are behind schedule and will work into the night. A crew of seven inmates are en route to assist."

"What are you talkin' about?" yelled Laris. "No one works the canyons at night. It ain't—"

Before he could finish, one of the nearby Ekron enforcers approached with bladed staff in hand, expertly twirling the weapon with the skill of a martial artist trained in the lost art of polearms. He struck Laris in the face with the non-bladed end of his staff, sending the inmate crashing to the ground. Marcus and his crew moved toward Laris to help, but the Androids raised their rifles toward them, forcing the crew to back off.

"You will carry out your orders as commanded," belted the Android. "Prepare the drill for operation."

As their captors backed off, Marcus moved in to help Laris to his feet.

"This is crazy," said Laris, attempting to shake off the blow to his head. "There's no tellin' what's out here."

"Calm down and grab your tools," said Marcus. "No one's ever seen anything down here," he said. "Let's knock this out so we can get out of here."

Laris reluctantly agreed and began prepping the drill alongside the rest of the crew.

Thirty minutes later, a group of seven miners working farther afield approached the base camp. While Marcus and his crew continued to work, one of the Ekron enforcers moved to meet the approaching miners.

As the enforcer passed a dark cave, three hideous bipedal creatures sprang from the dark recesses of the cavern, letting loose a cacophony of ear-splitting roars. The Ekron called the Martian beasts the *Zorgon*. The horned creatures stood at least twelve feet, with green scaly hides, and mouths full of jagged, spear-like teeth. They had golden-colored eyes with slit, multi-colored pupils capable of peering perfectly into the night.

With serrated claws, the Zorgons eviscerated the enforcer before he could react, prompting five of the seven approaching miners to bolt in the opposite direction. But they were immediately set upon by the trio of monsters and savagely torn apart. The two survivors sprinted toward the main base camp, seeking aid from the guards overseeing the drilling operation.

While one of the Zorgons continued to feast upon the flesh of the dead workers, the other two abandoned the carnage and gave chase to the two fleeing miners as they frantically sprinted toward Marcus and his team.

Though they walked on two legs, Zorgons ran on all fours, moving with speed and agility seldom seen in beasts of their size. Having no weapons with which to defend themselves, Marcus and his crew made a run for the nearby Mk17 storage crates, hoping to conceal themselves until it was safe to move.

During the commotion, the beasts tore one of the two miners in half before he could reach the base camp. As the lone survivor finally arrived, the Androids moved forward, opening up on the two monsters with their pulse rifles. The guards killed one beast, though it took a considerable amount of shots to do so.

Alerted by the rifle blasts echoing throughout the canyon, the Zorgon still feasting upon the flesh of the initial five miners abandoned its dinner to join the remaining beast still terrorizing the base camp. The ASUs did all they could to blast the monsters, but they moved with such speed that the Androids' advanced targeting optics had difficulties tracking them.

The two remaining Ekron enforcers used their bladed staves to ward off one of the Zorgons while the Androids assailed the other. But the monsters' ferocious attacks overwhelmed the enforcers, though the Ekrons managed to critically wound the animals before meeting their demise.

Despite their horrific injuries, the two creatures rushed the androids with extraordinary speed. The Zorgons toppled the robotic guards, ripping through their metal bodies like tinfoil.

During the attack, Marcus and his crew made a run for the bladed staves of the slain Ekron enforcers. As they approached the weapons, Marcus noticed the pulse rifle of one of the destroyed androids located far to the right of the animals. Marcus turned to Laris.

"I'm goin' for the rifle," yelled Marcus. "Distract 'em."

Laris and the other miners ran while yelling at the top of their lungs, trying to divert the nocturnal predators. The plan worked.

The creatures set their sights upon the defiant miners and approached, though much slower this time. It appeared the effects of their injuries were finally catching up to them.

"Hurry, Marcus," yelled Laris as he and the other inmates began swinging and jabbing at the creatures with their weapons, hoping to keep the exhausted animals from rushing.

Marcus made a heart-pounding dash toward the rifle. He pried loose the weapon from the Android's severed hand and opened up on the creatures from behind, dropping one of them.

While Laris and his crew moved in to slay the beast now flailing on the ground, the remaining Zorgon summoned the last of its strength to rush toward Marcus. But he poured on the rifle attack, despite the creature narrowly dodging the onslaught of crimson-colored energy bolts whizzing past its body.

As the Zorgon moved closer, the monster used its massive forearms to cover its face. And with a sudden force, the beast backhanded Marcus, sending him flying several feet back, causing him to crash hard into the ground, the impact separating him from his rifle. Marcus tried to move, but his entire body ached, feeling as if a hover-bus had just smashed into him, twice. But he counted himself lucky, as the creature's claws only partially caught him, resulting in bloody, horizontal gashes across his abdomen.

Before Marcus could stagger to his feet, the beast was upon him, releasing yet another bone-rattling, foul-smelling roar directly into his face. As the Zorgon raised its claws to kill Marcus, an inmate came out of nowhere from behind, grabbing the rifle from the ground to finish what Marcus started.

Firing off shots in rapid succession, the purple-haired inmate pressed forward, releasing energy rounds into the beast's hardened flesh until it at last fell to the ground, releasing its final breath into the chilly air of the Labyrinth of the Night. The inmate turned toward Marcus, reaching out his hand.

"I don't know how you survived out here without me," said the familiar voice.

Marcus looked up to see a welcome sight he hadn't seen in over four years. It was his former cellmate from Runner's End.

"Jackson," said Marcus in total surprise. "What are you doin' out here?"

Jackson turned his head to see the other inmates heading toward them. He turned back to Marcus and helped him to his feet.

"I was with the other mining crew, workin' a few miles up, when we were ordered to come here."

"I mean, what are you doing *here* at Noctis?" asked Marcus.

"To be honest, I have no idea," said Jackson. "About two days ago, shortly after lights out, I was dragged from my bunk by the guards. Next thing I know, I'm on a transport headed for Noctis."

While Jackson was well known for his over-exaggerated tales of adventure, tragedy and woe, Marcus could see no lie in his eyes. In fact, he saw only the look of confusion upon the face of his friend.

"Doesn't matter," said Marcus. "It's great seein' you again."

"Same here, man," said Jackson. "So, I guess this makes us even, huh?"

"More than even. You really saved my—"

Before Marcus could finish his sentence, an armored personnel transport descended from above, shining its spotlight on the inmates below.

"Guess they finally got the distress call from their buddies over there," said Marcus, motioning toward the mangled scrapheap that was once the ASUs.

"Yeah, and they took their sweet time gettin' here too," replied Jackson after dropping his rifle to the ground.

"Inmates. Drop your weapons and put your hands above your heads. Failure to comply will result in the use of deadly force," announced the ASU manning the heavy-assault cannon jutting from the side of the flying transport.

"That's our ride, boys and girls," said Marcus to the remaining inmates while nursing the painful wounds across his torso. "Drop 'em to the deck so we can get the hell outta here."

The inmates complied and dropped their weapons, as no one was interested in tangling with the armed transport hovering above their heads.

Moments later, the transport landed, rounding up Marcus and the rest of the survivors. As the ship finally pulled away from the valley floor, Marcus began to think as he looked over the aftermath of their deadly encounter, *Man, these last two months can't end fast enough.*

CHAPTER TWENTY

Over the next rotation, which was about the length of five standard Earth weeks, Marcus' stay at the Noctis work camps seemed a little more bearable since reuniting with his former cellmate and friend, Jackson.

Even Charles noticed a change in Marcus' attitude during their sporadic counseling sessions. But the circumstances under which his client's friend was transferred to Noctis concerned him deeply, as there wasn't even an official disciplinary hearing on file for Jackson. But Charles turned his attention back to Marcus.

Though the sessions never occurred as often as the warden promised, Marcus looked forward to speaking with Charles, as he often delivered to him news of his family back home.

"You only have a few weeks left," said Charles from the small view screen in the claustrophobic storage room they used for their remote sessions.

"Yeah," said Marcus, "but I need you to get Jackson out of here too," he continued. There was an uneasy feeling regarding his friend eating away at the back of his mind. "How he got here, it just don't feel right, you know?"

"Don't worry," said Charles. "I have people on Earth looking into it as we speak," he continued. "Just make sure you stay on the straight and narrow until you get out of there."

Before Marcus could respond, one of the ASUs burst into the room.

"Your session is over. Report to your mining detail immediately," said the android.

Marcus still had ten minutes remaining in his session and started to protest. But Charles interrupted before Marcus could say something, he might later regret.

"It's okay, Marcus. I was going to end it here anyway. Handle your business. We'll talk later."

Marcus sighed, but kept his mouth shut. He followed the guard out of the storage room without protest.

Hours later, Marcus found himself back on the valley floor, in the same location where they had encountered the ferocious beasts just one month earlier. *At least they beefed up security*, Marcus thought, though he felt the increased guard presence was more for protecting the equipment than for the miners.

For the next five hours, the gargantuan mining drill, manned by six inmates, ripped through the canyon wall with its pulsating green energy beam, melting the rock away, exposing the reddish-black chunks of Krillium ore hiding behind the thick rock face.

The miners manually collected the smaller pockets of ore too miniscule for the drill's extractor to retrieve, using handheld battery-operated laser drills equipped with portable ore extractors.

Marcus was hard at work next to Laris, chipping away at a stubborn piece of rock, when the beam of his drill started to flicker just before shutting down.

"I'm out," said Marcus. "Where'd they move the battery packs? I didn't see them in the usual spot."

"They're in bunker twelve, at the edge of camp," replied Laris.

"Why'd they move them all the way out there?"

"I stopped tryin' to figure this place out a long time ago," said Laris. "Want me to grab one?"

"No, I got it," said Marcus, "I need to take a piss, anyway."

"Didn't need to know that," said Laris as he continued to melt the loose rock away with his hand drill.

Marcus chuckled and headed toward the cave's entrance. And after making a quick stop behind a large boulder to answer nature's call, Marcus made the half-mile trip to the edge of camp toward bunker twelve.

He arrived at the makeshift storage warehouse to find it a complete mess, full of unmarked storage containers and all manner of mining equipment strewn throughout the area. Not even knowing where to start, Marcus rummaged through the shelves hoping to stumble upon the battery packs. But he came up empty.

Before moving to the next shelf, he was attacked from behind by a large, menacing figure of great strength. He never could maneuver himself to see the face of his attacker, but one look at those massive forearms of skin and living metal and Marcus knew his attacker to be a Gorean Cyborg. He was certain of it, having sparred enough in the past with Tony, his former Gorean crewmate, to recognize that unique physiology anywhere.

Marcus used his legs to push off the walls, hoping to bring him and his attacker to the ground, perhaps breaking the hold altogether. But the Gorean's crushing bear hug remained unbroken. He continued to struggle, when a second masked assailant, much smaller than his current attacker, approached from his left.

Moments later, Marcus felt a painful piercing sensation in the side of his neck. Everything in the room went black.

Marcus later awakened to find himself covered in blood, though thankfully not his own. But to his dismay, he found himself among the mutilated remains of his friend and former cellmate.

"Jackson!" yelled Marcus amid feelings of anger and deep remorse.

To the left of Jackson's body, Marcus could see the murder weapon, a two-foot pipe with one end filed to a point, perfect for piercing and slashing attacks, a fate to which his friend had so obviously succumbed. Enraged, Marcus tried to stand, but from his perspective, the room seemed to spin out of control. Then the debilitating headache followed, causing him to stumble to the ground.

Moments later, the ASUs arrived as if on cue, no doubt in response to an anonymous tip from the true mastermind of that heinous crime. And before Marcus could even deny his involvement, an android struck him in the head with the butt of a rifle, causing the room to once again go dark.

#

The investigation that followed was a total farce. Were it not for his counselor Charles Harris' extensive contacts, Marcus might have found himself up for execution, or perhaps a more permanent stay at Runner's End.

Charles had a special investigator shuttled in from Earth, tasked to review the circumstances of the murder. The investigator's findings ultimately contradicted the results of the warden's so-called *internal investigation*. And after the independent review was complete, there were enough discrepancies found to keep the warden from proving beyond a reasonable doubt, a standard still held in human legal systems of the 32nd century for murder convictions, that Marcus was the killer.

While Marcus had dodged the worst of that bullet, the warden began hatching the next phase of his plan, designed to further add to Marcus' suffering. Ian argued that given his five-year disciplinary track record, Marcus was both a danger to himself and the other inmates and should therefore be placed into stasis for the rest of his prison term.

While Charles' efforts kept Marcus from the execution chambers, he couldn't stop what was coming next. And thanks to the further manipulation of Jazmin behind the scenes, Ian got just what he wanted. With another unanimous vote from the disciplinary board, they sentenced Marcus to the dreaded stasis chambers of Runner's End for the remaining two-and-a-half years of his sentence.

CHAPTER TWENTY-ONE

It was like drowning beneath the ocean waves, yet never running out of oxygen. Every labored breath was like some small victory in a never-ending struggle between Marcus and the stasis chamber's energy field in which he found himself encased.

It wasn't supposed to be like this, Marcus thought as he lay dormant, completely conscious yet unable to utter a sound as that diabolical instrument of incessant torture restricted his vocal cords. While experiencing no physical trauma from the chamber itself, the mental anguish was almost too much to bear, making it impossible to keep his mind from wandering.

As he lay there, locked away in a dim vault, all he could do was reminisce on events that led him to that point. But he snapped out of it as the thought of what he endured only frustrated him further. He stared out of his stasis chamber into the poorly lit room containing thousands of chambers filled with other inmates. They were all neatly stacked away, out of sight and out of mind, which is what the vindictive Ian Trent did to those who were an *inconvenience* to him.

For the first time in his life, Marcus felt as if he had no control, not even of his own body. The warden had taken everything. *Or was it perhaps the choices he made in life that did this to him?* It was a point that Charles often argued during their tension-filled sessions throughout his incarceration. But at that point, he figured it didn't matter. *What's done is done.* Despite all the progress he had made to better himself over the years, not to mention the promises he made to his sister and niece, he felt as if the warden had won.

All he could do was close his eyes and sleep, wondering if it was all worth the effort, or if he would even survive his final years in stasis. As Marcus rested from all his toils within the confines of that contemptible prison, Ian Trent stood silently in the shadows with a searing gaze fixed on Marcus' stasis chamber.

Through five years of missteps, Ian felt as though he finally had Marcus right where he wanted. *After all, stasis technology is far from an exact science, and accidents happened all the time,* he thought, barely able to suppress the smile cracking the corner of his mouth.

"I wouldn't start that victory lap just yet," said Charles as he stormed into the room from behind.

"Mr. Harris, how nice of you to join me," said Ian. "This is for the best, you know," he continued in a disingenuous tone. "He'll be safe here."

"I know he will," said Charles, taking a long look at Marcus' stasis chamber.

Before he could say another word, Ian watched in total shock as an entire detachment of unknown human guards marched into the room. The new guards relieved Ian's personal Android Security Units of their duties.

"What the hell do you think you're doing?" yelled Ian, furious that Charles somehow locked him out from giving his ASUs orders. "This is *my* prison," Ian shouted. "How dare you—"

"Consider your ASUs relieved, Warden," said Charles as he produced executive orders signed by Jena Oxana, the longtime Planetary Governor of Mars.

Governor Oxana was a staunch supporter of the current ISL President, Jonathan Vance. She also had a strong disdain for candidate Ian Trent and everything for which the *Orion's Shield sympathizer* stood.

"The elections are less than a year away, Ian. And the Governor knows you are a busy man these days," he continued, enjoying the look on Ian's face. "She felt such a *high value* ISL prisoner deserved a few extra eyes watching him. So, she thought she'd help you out."

Ian started to object, but Charles cut him off.

"The Governor says, *you're welcome*, by the way," said Charles. "And she sends her warmest regards to you *and Jazmin*."

"You'll pay for this," said Ian.

"A fresh contingent of the Governor's personal guard will stand watch over the area every week, until Marcus is released," said Charles. "And if Marcus' chamber so much as experiences a power fluctuation during that time, I promise you this, warden, you *and your staff* will become permanent residents of Runner's End, courtesy of Governor Oxana, and yours truly."

Ian stormed from the room without saying a word. Charles smiled as he watched Ian exit the facility in ignominy. He then walked over to Marcus' stasis chamber, placing his hand on its transparent outer casing.

"You've been through a lot, son," said Charles. "Rest, and I'll see you on the other side."

Charles made an about-face turn and headed toward the exit, saluting the new commanding officer of the watch on his way out. As the automated door slid shut behind him, Charles continued down the walkway toward his office. Before entering his place of work, he looked back toward the stasis chamber area one last time and exhaled. *Don't worry, son,* Charles thought. *I'll get you home to your family.*

PART THREE

A CLASH OF DESTINIES

"You know what I do. You know what I'm about."

— *Daren West*

CHAPTER TWENTY-TWO

The wailing of alarms and the unmistakable sounds of weapons fire ringing out in the dark jarred Marcus awake. His heart raced, and adrenaline flowed, yet his body would not respond. In stasis, the only luxury of movement inmates could experience was the ability to open and close their eyes.

From across the way, he could see dozens of Android Security units deactivating stasis chambers, ripping open the doors, and executing the occupants. He tried desperately to regain some ounce of control of his physical faculties, but it was an exercise in futility. A short time later the stasis field in his chamber deactivated. The sudden release of the energy field left him weak and barely able to move. And with a powerful unseen force, the door of his chamber was pulled off its hinges.

Marcus looked up to see the silhouetted faces of two Android soldiers, with only their pulsating blue eyes piercing through the darkness. They stepped aside, giving way to a menacing individual clad in black. The mysterious figure moved closer to Marcus, finally revealing his identity. It was the warden, Ian Trent, staring at him with a baleful grin. Marcus' face hardened with anger. He tried to summon any remaining strength he had to charge at the warden, but Ian caught him by the throat before he could move. Ian yanked Marcus from the chamber and pinned him against a nearby metal column, his feet suspended several inches from the ground.

Stunned by his attacker's extraordinary strength and heightened reflexes, Marcus began to think. *Has Ian been augmented with some sort of Viribus Implant?* Then Marcus heard the warden speak, his tone sounding as if a multitude of voices were crying out in unison in pitches both high and low.

"I told you we'd come to know each other very well, Mr. La'Dek," Ian said in a surreal and portentous tone. "All deeds come with a price," he continued, echoing the ominous words spoken to Marcus by his former rival, Jarek Malasen. "Time to pay what you owe."

Marcus couldn't respond under Ian's crushing grip around his throat. He then considered the warden's eyes; they had a fiery red glow about them. The warden displayed a toothy grin with teeth that resembled jagged stalactites and stalagmites. As his tormentor belted out a disturbing laugh, Marcus looked down at Ian's right hand. The warden held the very serrated thermo-blade that Jarek used against Marcus years ago, its edge glowing red hot.

Was this finally it? Marcus thought. He had accepted long ago that everyone, including himself, had to punch out at some point. Marcus just never envisioned being gutted by a spineless coward like Ian Trent, left to bleed out on a dirty floor of metal and stone, with no one to see, or even care for that matter.

And with one ferocious thrust, the warden impaled Marcus' abdomen with his weapon. He groaned in agony as Ian literally twisted the knife. Everything around Marcus began to grow dim. Marcus felt once again as if he were drowning. But this time he truly *was* running out of air, and out of time. Hyperventilating, Marcus tried desperately to clutch the last vestiges of a life rapidly slipping away, and then...

#

Marcus sprang awake in his bunk, disoriented and gasping for air. It was another nightmare; one of many he'd experienced since his time as an inmate of Runner's End for nearly a decade. By Interstellar League time, which followed the standard Earth calendar, it was *the year 3120*—just over thirteen months, or one full cycle, since his release from prison.

They said that Marcus had paid his debt to society, but the final two and a half years he spent in stasis had nearly broken him. No human prisoner had ever been subjected to such an extended period within the confines of those vile chambers. The doctors said the nightmares, most of which he experienced during his time in stasis, were side effects of both the sedatives used, and the energy field emitted by the chamber itself. He was told the condition would pass in six to eight weeks, but over a year later his dreams were still beleaguered by the disturbing imagery, often robbing him of his beloved sleep.

Marcus lay back in his bunk to find his pillow drenched with sweat. He tossed the pillow to the deck and rested his head directly on the uncomfortable mattress. He stared at the ceiling of the cramped, metallic crew quarters of the industrial-class cargo freighter he'd been piloting for the last two days. The ship proudly displayed the faded markings of *La'Dek Transports* across its battered hull.

The freighter had been traveling silently through deep space at barely faster than light-speed thanks to outdated, barely functional Hyper Jump Engines that should have been retired ages ago. His destination was the Interstellar Star Port, Proteus, where he was scheduled to deliver a payload of satellite components and foodstuffs to local businesses, barely scraping by in those economically trying times.

The ship with which Marcus had been fighting to keep operational for the last forty-eight hours was a small Lambda-class hauler, a designation given to light-cargo freighters. But despite its age and overall condition, the ship performed well enough for small cargo runs. However, his vessel's performance was eclipsed by the colossal Omega-class freighters used by the competition, which far exceeded his craft in size, speed, and cargo capacity.

His family's fifty-five-year-old Cargo Shipping Company started by his late father, William La'Dek, was now run by his older sister, Samantha. What was once hailed as *the galaxy's premier choice for all of one's cargo shipping needs*, La'Dek Transports had been reduced to something of a discount cargo hauler, making runs for financially strapped businesses that were no doubt struggling as much as their shipping company.

While ever thankful for the employment opportunity provided to him by his sister, Marcus absolutely hated his job. But it was the only honest work he could find. There weren't many businesses eager to hire an ex-con who'd never held a *real job* his entire adult life. And if employers knew the details of his former *occupation*, he'd never be allowed through the door, much less offered gainful employment.

Marcus activated an ISL newsfeed on a nearby monitor as he stood to his feet to stretch. He rubbed the still visible scars on his abdomen—courtesy of the dreadful beast he had encountered at the Noctis work camps—as his mind briefly drifted back to darker days. He turned toward the reflective surface of the bulkhead next to his bunk and sighed. He stared grimly at the scar caused by the sniper round he had taken to the left shoulder on Titan, an old wound that served as a dour reminder of a failed life that ended eleven years earlier.

As painful memories bubbled to the surface, Marcus shifted his attention to Bobby Wiseman, the fast-talking, silver-haired reporter rambling in the background on the monitor behind him.

"Richard Maxis, CEO of Interplanetary Defense contractor Maxis Corp, is due in court today amid allegations of treason," said Wiseman in his trademark, melodramatic tone. "The high-profile businessman is charged with conspiring to sell critical prototype technology that could very well be the answer to the galaxy's Terraformer energy crisis to Teric Winters, leader of the Outer Core Terrorist Faction, Orion's Shield."

Marcus shook his head in disbelief, wondering why an ultra-rich business tycoon such as Maxis would even bother with the likes of Teric Winters and his fanatical band of sociopaths, of which Marcus himself was a former member and knew all too well.

"But in a bizarre twist to this ongoing scandal," continued Bobby Wiseman, "reports are beginning to surface that the prototype may have been stolen from Maxis Labs prior to the ISL government moving in to seize the sensitive technology. The surprise theft was alleged to have happened as many as seven weeks ago.

"Thus, tensions continue to escalate between the ISL government and the Lyrian Empire, setting the two civilizations on a collision course that some experts say could end in all-out war," Wiseman continued in a tone both dark and foreboding. "A war that many say the human-led ISL faction would have little hope of winning."

Marcus became exceedingly angry as he listened to the story that had been dominating the news cycle week after week.

"ISL President Jonathan Vance is set to address the council of the Galactic Planetary Consortium in the coming days—"

Marcus abruptly terminated the news feed, as the entire story irritated him. Ever since news broke about the Maxis Corp scandal, business had slowed considerably for La'Dek Transports, mainly due to the proximity to which their company operated near Lyrian Space. At the end of the day, it wasn't just the news report that bothered him; it was everything about his life.

And now in the lonely confines of the small cargo ship, flying silently through the vast loneliness of space, Marcus struggled with his new reality, a life where he'd likely grow old and die broke, leaving him to wonder just how long he could continue to walk this straight and narrow path.

CHAPTER TWENTY-THREE

"Incoming transmission from Samantha Brown-La'Dek," announced the monotone digital voice blaring through the ship's speakers.

Marcus headed for the bridge to take the call. It was a cramped area packed with the ship's flight controls, navigation computer and other electronic accouterments necessary for the vessel's operation. The main view screen, which typically displayed the area of space directly off the ship's bow, was now overlaid with the real-time image of his sister, Samantha. She was broadcasting from her small two-bedroom apartment on the *Stratus One Star Port*, located on the edge of ISL-controlled space.

A gorgeous forty-nine-year-old widow, often mistaken for a woman in her mid-twenties, Samantha took remarkable care of herself, a trait picked up from the ten years she spent serving in the officer ranks of the Interstellar Guard on Earth. Marcus always suspected Samantha was more than a lowly logistics officer. But he never could get her to disclose the true nature of her duties while she served in the military.

"Is he on yet?" asked Elizabeth, Marcus' seventeen-year-old niece.

Before Samantha could respond, Elizabeth rushed to the camera, almost knocking her mother from her chair.

"Happy Birthday!" Samantha and Elizabeth yelled in unison.

Marcus tried to maintain his tough-guy demeanor but couldn't help cracking a smile.

"And how old are we today?" asked Elizabeth.

"Too old," he responded.

"Calm down, boy. Thirty-nine is nothin'. Besides, we age well in our family," Samantha said, flipping her long dark hair to the side. "Just look at me."

"You call that aging well?" Marcus asked, taking a cheap shot at his older sister. But she let it slide in honor of his birthday.

"Don't forget, we're supposed to tour the universities on Earth when you get back," Elizabeth chimed in, knowing the promise had already slipped her uncle's mind.

"How could I forget?" he said, trying to play it off.

"Well, I have to get ready for school. I love you and I'll see ya when you get back."

"Love you too," Marcus replied.

As she walked away, Marcus could see she was in a great deal of pain. He noticed Samantha watching her as well, seeing the heaviness on her face. But before he could speak, Samantha changed the subject.

"You know, Charles has been trying to reach you," said Samantha. "Said he hasn't heard from you in months."

"I see the two of you have been gettin' close," Marcus said, attempting to redirect the conversation. "Somethin' you wanna tell me?"

"First of all, stay out of grown folk's business," Samantha said jokingly. But she wasn't about to let him off the hook. "I'm sure he just wants to know you're alright," she continued. "After all, you've been through a lot..."

"I'm fine," Marcus said with a slightly elevated tone. Besides, he didn't want Charles seeing him like this, given all the counselor did to get him back home to his family.

Samantha could see he had no desire to talk about it, so she backed off. The two sat silently for a moment.

"How's Liz doin'?" asked Marcus, finally breaking the silence.

"You know, she has her good days and bad days," Samantha replied after taking a deep breath.

"What'd the doctor say?"

"It's terminal," said Samantha following a long pause. It took everything within her to keep from breaking down.

Marcus' head dropped. He felt as if the wind had been knocked out of him as he too tried to hold it together. "Are they sure?"

"They have no idea what's causing her condition," Samantha said. "It's some form of degenerative disease, affecting her muscles and bones. And it seems to worsen the older she gets."

"There has to be something they can do," said Marcus, raising his voice.

His sister took no offense to his tone. She let him vent, silently listening as he went on for several minutes, blasting everything from the useless doctors that he felt were bleeding them dry, to the subpar medical facilities on that run-down Star Port on which they lived.

"It's not all bad," she said, trying to cling to some residue of hope. "Liz is responding well to the experimental treatment. They think we may be able to manage the condition, hopefully extending her life well into adulthood."

"And the catch?" asked Marcus. "There's always a catch."

"The price tag," Samantha admitted. "I'll be straight with you. At this rate, the cost of her long-term care is gonna bankrupt us and the business if we don't close more contracts," she continued. "I'm working with the banks to get another loan to upgrade and expand our fleet."

"Are they gonna do it?"

"Between these ridiculous GPC trade restrictions and all this talk of a possible Lyrian invasion in the news, people are afraid to do business out here. The loan officer thinks we'll be unable to raise the capital needed to repay—"

"Bet he'll change his tune if I come down there and shove my rifle up his—"

"Stop it, Marcus," said Samantha. "It doesn't work like that, and you know it."

"You play by the rules, you get burned every time," said Marcus, rubbing the back of his bald head. "I'll come up with something, I promise you that," Marcus assured his sister.

But she knew what he was thinking. "Don't go out there and do something stupid, Marcus," she said. "You just got out and we're *not* about to lose you again."

"Whatever you say, Sam," replied Marcus, knowing Samantha's convictions would never allow her to see things his way.

Samantha could see the frustration building, so she switched topics. "I'm impressed; it looks like you're actually ahead of schedule for your cargo run to Proteus."

"Yeah, no thanks to these outdated hyper-jump engines. Dad should've scrapped these things years ago."

"C'mon, Marcus. Do we really have to do this again?" Samantha asked, sensing another *blame it on Dad* conversation coming. "He may not have been candidate for father of the year, but he did the best he could with what he had."

"The best he could?" said Marcus, surprised she let those words part her lips. "With all the fines he racked up over the years behind his shady business deals, it's a wonder we're still in operation."

"Says the guy that just spent ten years in prison for his own *shady* dealings," Samantha countered quickly.

She had him there and he was furious, but he held his tongue.

"After Mom walked out on us, the decisions he made ravaged this company and our family, I'll give you that," Samantha said. "But he changed, Marcus. You just weren't around to see it."

"Yet here we are, stuck cleanin' his mess at a time when we should be—"

"Stuck?" Samantha interrupted. "There are humans out there sellin' themselves as slaves just to feed their families. Last I checked that wasn't our story. What's with you?"

"Look at me, Sam," he said. "I'm nearly forty years old, broke, and livin' with my sister. Not exactly how I envisioned my life, you know?"

"Yeah, Marcus, I do know. You're not the only one who had dreams," Samantha responded. "I gave up everything to keep this business afloat while you were out roaming the galaxy, playin' the big bad pirate," she continued sharply. "So, you don't get to sit there and complain..." She stopped herself, seeing the conversation starting to spiral out of control.

While he wouldn't have allowed anyone else to speak to him in that manner, Marcus often held his peace when it came to his sister, even though she would, at times, speak to him as if he were her child. But he didn't mind. After all, not only was there a ten-year age gap between them, but Samantha did also have to largely step into their mother's shoes when they were kids after their mom ran off with that space pirate. So, Marcus often took no offense to the motherly tone of his sister. Besides, not only was she almost always right in these matters, he relied on her to be that steady voice of reason, helping to keep him on track when everything within him wanted to veer off course.

"Look, I'm proud of you," said Samantha, once again changing the subject. "Startin' over ain't easy. But I believe God has something better for us down the road. We just have to stay the course."

The navigational computer beeped, notifying him that he was approaching his destination.

"Well, this is my stop," said Marcus, rubbing his slightly graying beard. "And thank you once again for another *engaging* conversation."

"Hey, that's what I'm here for," she said. "Who else is gonna save you from your stupidity?"

Marcus laughed. "And let's hope the Government finds that missin' prototype so they can squash all this war talk. Maybe we can land a few more contracts."

"We can only go up from here, right?" Samantha asked optimistically. "I love you, and we'll see you when you get back."

CHAPTER TWENTY-FOUR

The weather called for clear skies, though the chill in the air refused to yield to the warmth of the penetrating sun. The sky traffic was unusually light around Earth's Haven City. Many had taken the week off in observance of the founding of the Interstellar League of Planets. However, in the heart of Haven City, inside the immense fortress known as *The Citadel*, there would be no rest for the weary.

For hours, ISL President Jonathan Vance had been locked inside the inner room of the presidential office known as the Virtual Council Chamber—affectionately called *The Box* among Citadel staffers. He was attending an emergency council session with the other faction leaders of the Galactic Planetary Consortium—known as the GPC—of which the ISL was a fledgling member. *The Box* featured advanced holographic emitters that projected a solid, three-dimensional replica of the council chamber found on Korax Prime, the planet where the GPC was founded more than five hundred thousand years ago.

The chamber was a large, lavishly decorated amphitheater featuring tiered, bench-style seating of stone. Each seat cushion was embroidered with ornate symbols etched upon exotic fabric from across the galaxy. Surrounding the oblong-shaped speaking platform on the lower level were jewel-encrusted thrones upon which the rulers of the ten prominent races that comprised the GPC sat, serving as representatives for their people.

The Consortium's primary function was to create, maintain, and enforce interstellar law within the known sectors of the Milky Way. The GPC also regulated trade and commerce among its member races. Unfortunately for President Vance and his fellow humans of the Interstellar League faction, power was not equally shared among the GPC's member races. This was especially true when it came to the Consortium's largest, most influential member, The Lyrian Empire.

Lyrians were a proud humanoid race, almost reptilian in appearance, except they had long, flowing hair like that of the majestic mane of a lion. They had heavily scaled faces of varying tones, as well as jagged, razor-sharp teeth. Despite the Lyrians' menacing appearance, they were an elegant, well-spoken race of cultured warriors, believing themselves to be descendants of dragons—the mythical creatures spoken of throughout many of Earth's ancient legends. Because of these self-proclaimed origins, many among their kind believed themselves to be the true heirs of planet Earth, causing much division between them and the primitive human race of the ISL.

In that moment, the Lyrian Emperor, Lord Zek'Ren, dominated the conversation on the council room floor as he spoke before a captive audience that was hanging on his every word. He was a tall yet stocky Lyrian man of considerable girth. He wore a ceremonial tunic with matching pants and boots in varying shades of purple and black. The entire outfit featured intricately designed padding, not unlike the scales of the dragons depicted on the many tapestries that adorned the grand halls found on their home world, Lyria Prime.

Zek'Ren also wore metallic, jewel-laced gauntlets that covered his muscular forearms. The entire outfit was slightly covered by a royal blue, floor-length hooded cape attached with circular, chain-linked clasps that were shaped to look like bucklers. Upon the clasps were extravagant, luminous engravings resembling dragon fire. The clasps seemed to glow red hot as his voice modulated.

"Never in the Consortium's 500,000 Full-Cycle history has such an egregious act been committed by a ruling faction," Zek'Ren said as he addressed the packed council chamber.

Most of the attendees, like President Vance, attended the meeting virtually, their holographic images projected directly into the chamber seats.

Zek'Ren turned to his left to face President Vance. "What have you to say for your people?"

President Vance stood to his feet, relieved he could finally get a word in following Lord Zek'Ren's marathon speech. Without making eye contact with the arrogant Lyrian, he turned to address the other eight faction leaders on the council chamber floor.

"I can assure everyone on this council we're doing everything within our power to recover the stolen prototype—"

"Funny how your own people seem to conspire against you," said Zek'Ren, rudely interrupting. "Perhaps we *should* reconsider the Interstellar League's membership within the Galactic Consortium?"

"With all due respect, Lord Zek'Ren, it's not just *my people* we're dealing with here. Our intelligence reports show that even Ar'Gallious has been after the prototype for months," the President continued. "So, it appears you also have a dog off its leash."

At that moment, the entire council chamber fell silent.

"Ar'Gallious is no longer part of the Lyrian Empire. He and his so called "Dragen Alliance" are confined to the Outer Core star systems."

"Well, if you'd bothered to actually *recon* the O.C. from time to time, you'd know he's been amassing forces for decades," Vance fired back. "Most of which are defectors from your own Empire," he said. Before Zek'Ren could respond, Vance digressed. "But no, you're too busy crafting unfair economic policies, forcing many of my people to sell themselves as slaves just to feed their families."

"Know your place, human." Zek'Ren roared before the council chamber, bringing an end to the President's short-lived tirade.

President Vance looked about the room. He witnessed many humans working as indentured servants and slaves for many of the Xeno-dignitaries. His heart sank as he watched one of Zek'Ren's own assistants beat a human slave for nearly spilling a drink. But before Vance could further address the atrocities visited upon his people, Lord Zek'Ren interjected.

"You were required by the same scientific mandate set forth to all member races of the GPC," said the emperor, "to surrender all research intended to solve the Terraformer energy crisis plaguing us all." Zek'Ren turned his back to Vance. "And now, you would have us to believe that your self-proclaimed answer to this century-old problem has suddenly gone missing?"

"I've told you already, we're tracking the Orion's Shield operatives we believe to be responsible for its theft."

Emperor Zek'Ren turned to face President Vance. "Perhaps it is time to, as your people say, *get your house in order*, Mr. President?" said Emperor Zek'Ren. "Orion's Shield is your problem. So, let me put it in terms that even a human can understand." The emperor again turned his back to President Vance. "You have one galactic cycle to find that *conveniently* missing prototype, or risk expulsion from the GPC," Zek'Ren warned, his boisterous voice filling the room.

"One cycle?" shouted President Vance. "We're not searchin' just the Sol system," he continued. President Vance was furious, for one galactic cycle was slightly less than seven months of standard Earth time. He felt a full-cycle, which equated to just over a year, would have been more reasonable. "Our projected search radius spans numerous star systems from here to the Outer Core! It would take at least a *full cycle* just to—"

"Silence, human!" Zek'Ren roared. "The logistics of your investigation are not our concern," said the Lyrian, coldly. "We are interested only in results. One cycle, Mr. President, no more." A momentary icy silence fell over the room. Zek'Ren moved face to face with President Vance, then spoke. "Should you fail, I promise you this, we *will* cross your galactic borders and rectify the situation ourselves."

President Vance would have loved nothing more than to choke the very life out of that egotistical tyrant. And perhaps he would have, were it not for the fact that his hands would simply pass through the Emperor's virtual image.

Upon termination of the emergency session, the entire council chamber vanished as the holographic emitters powered down. With a sigh of frustration, President Vance exited *The Box* and headed toward his office. The stench of war was in the air. And it was coming at a time when the ISL could least afford a sustained conflict. As he strode the carpeted hallway of the Citadel, he gazed upon the hand-painted portraits of the ISL Presidents that served before him, wondering how his predecessors held it together in such precarious times. He was just three years into his second presidential term, after narrowly defeating his political rival, Ian Trent, the infamous warden of Runner's End. And while ever grateful to once again take up the mantle as the Interstellar League president, it was times like these that made him regret ever running for a second term. He reached the mahogany door of his office and stepped inside. It was time to assemble his top advisors. And they would need to act quickly, lest peace was to become a luxury they would no longer enjoy.

CHAPTER TWENTY-FIVE

Following his lengthy cargo run to Proteus, Marcus finally returned home to the Stratus One Star Port on the remote edge of ISL-controlled space. It was an immense circular superstructure, resembling a giant twelve-spoked wheel rotating in space. While serving as a major trading hub between the Lyrian Empire and the ISL, Stratus One was often frequented by smugglers, thieves and pirates looking to move illicit cargo between empires, mainly due to its inadequate security forces policing the day-to-day affairs of the station. Marcus despised nearly everything about the place. It was an ever-present reminder of a tumultuous childhood he tried desperately to leave behind. To him, Stratus One was little more than a soul-sucking ghetto floating in space, where good, well-intentioned dreams go to die a quiet death in the cold vacuum of space.

After docking with the station, he gave the awaiting deckhands instructions regarding the handling of his cargo. They were mostly teenagers in tattered clothes, working hard to put a few scraps of food on the table for their families. Marcus tipped what he could, but times were hard for everyone. They eagerly accepted the money and ran off to complete their assignments.

Upon exiting the docking bay, he entered the expansive circular section of the star port, which resembled an Industrial sector of a large city. The area was filled with buildings, both great and small, occupied with businesses ranging from street-level farmers' markets to enormous industrial buildings, housing commercial manufacturers that produced mainly satellite and starship components. The industrial and commercial sectors of the star port were divided by a large two-lane highway that ran the circumference of the space station. Around the clock, a seemingly endless stream of personal and public transports traversed the highways, hovering several inches above its grimy metallic surface.

Instead of taking a nearby public transport, Marcus opted to walk the half-mile trip back to the office of La'Dek Transports where Samantha, Liz and their cousin Kayla were no doubt awaiting his arrival. As he made his way down the walkway, he narrowly dodged endless crowds of pedestrians and hordes of would-be entrepreneurs looking to peddle mostly illegal wares to those unfortunate enough to cross their paths.

Operating on the outer fringes of civilized space, the problematic *flesh trade* was a major issue on Stratus One. Everywhere he looked, Marcus could see zeth-addicted men and women of varying species offering their bodies to the highest bidders at the behest of their handlers.

But these so-called *handlers* were little more than shady-looking slave drivers who had many of the local security forces in their back pockets, allowing them to operate with almost total impunity.

Marcus covered the distance to the office in under ten minutes, feeling as if he needed to take a shower after the experience. Before entering, he grabbed a bite to eat from a neighboring food vendor. As much as he hated Stratus One, *at least the food is good*, he thought as he entered the small office building of his family's cargo shipping company. It was a small workspace they shared with three other businesses operating on the upper levels. The layout was nothing special, just five desks with computer terminals and two back offices, one of which belonged exclusively to Samantha, completely off limits to everyone.

As Marcus entered the door, he was immediately rushed by Elizabeth. She took after her mother, though she was a bit lighter in complexion. However, her defining physical trait was one that both baffled doctors and enhanced her beauty, emerald green eyes that were almost unnatural in appearance. She squeezed her uncle tight, talking his ear off about every stop she wanted to make during their upcoming trip to Earth.

As Elizabeth yammered on, her older cousin Kayla approached. She was a tall, athletic college grad. She moved to Stratus One from Earth four years ago to help Samantha rebuild the family business. She saw it as the perfect opportunity to gain some much-needed experience, hoping to one day become as successful a businesswoman as her cousin, Samantha.

"Good to have you back, cuz," said Kayla. She moved closer. "Your sister has been on one since you left, so watch out," she whispered into his ear.

"Thanks for the heads-up," said Marcus.

"So, when are we off to Earth?" asked Elizabeth.

"And hello to you too, Liz," said Marcus while prying his niece's arms from around his torso. "I see you're feeling better today."

"Yeah, I started the new treatment last week, and I think it's finally kickin' in. I feel like I can run around this entire station," she said.

"Not without me, or a heavily armed escort, by your side," said Marcus. And he was serious.

He would never let his niece walk down the street by herself on that disgusting, crime-ridden space station. While Elizabeth may possess a beyond-genius-level intellect, complete with an uncanny ability to learn and speak any language, she had no idea what life was like outside the sheltered confines of her mother's home.

"So, when do we leave?" she asked again. "I'm all packed."

"Would you give your uncle a chance to breathe, girl?" said Samantha to her over-anxious daughter. "He just got back," she said while shoving a stack of data pads into her brother's chest. "These are the latest contracts," she said to Marcus after giving him a quick kiss on the cheek. "Welcome home."

"What happened to getting a chance to breathe?" asked Marcus as he cradled the stack of data pads in his arms.

"You know this is our busiest time of the year," she said, referring to harvest time on Phaelon Six, the lush green planet over which Stratus One was orbiting. "The farmers will be here soon. You can rest when you're dead," she said to Marcus while ushering him to his desk.

"And you two," Samantha said to Elizabeth and Kayla, "quit playin' around and get back to work. I need those invoices and manifests completed yesterday."

"Yes, drill sergeant," replied Elizabeth while giving a halfhearted salute. But one look into her mother's eyes, and she thought it best to join her cousin to resume their duties.

Marcus could see she was in full military mode and opted to start on his own avalanche of data pads.

Moments later Samantha returned to his desk with a cup of coffee.

"Sorry for the less than *warm* welcome," said Samantha while handing over the small peace offering. "I'm glad you're home," she said while taking a seat on Marcus' desk. "Did Mr. Johnson give you any crap on Proteus?"

"You know he did," replied Marcus. He then reached inside his duffel bag and retrieved a small data pad. "But the deal is sealed," he said as he handed the signed contract over to Samantha, who was now grinning ear to ear. "That cheap bastard wasn't gonna find a better deal to deliver his jump-coils."

"You're the best," said Samantha while giving her brother a smothering bear hug. "We needed this one."

"Yeah, but don't you think we're kinda idling our thrusters here?" asked Marcus, lowering his tone slightly so Elizabeth wouldn't hear. "I mean, we're gonna need a lot more than this to keep up those treatments and make a living in the process." He then spoke just above a whisper. "I might know a guy..."

"Is it legal, Marcus?" asked Samantha, already knowing where her brother was headed.

"What kind of question is that?" replied Marcus. "I mean, it's legal... *ish.*"

"C'mon, Marcus. We've been through this," Samantha said amid a sigh of frustration. "We do this right, or—"

"Not at all." said Marcus, completing the sentence he'd heard a thousand times. "How much longer do you think we can keep this up?" asked Marcus.

"As long as we can," Samantha replied as she watched her daughter enjoy a rare day of almost perfect health. Marcus knew there was no arguing with her. But he had no intentions of sitting idly by, watching his niece waste away. So, he resolved within himself to do whatever it took, even if it meant breaking promises he'd made to himself and his family.

CHAPTER TWENTY-SIX

"Thank you for cutting short your well-deserved time off with your families," President Vance said to his two chief advisors seated at the cherry oak table in his personal office.

First, there was ISL Intelligence Taskforce Chief, Sarah Miller. She was a stern, meticulously organized redhead in her mid-forties, whom had been proudly serving as intelligence chief for the last six years. Next to Sarah sat the newly appointed Supreme Commander of the Interstellar Guard, General Thomas A. Kirkland. The fifty-five-year-old general's appointment to Supreme Commander was thanks in no small part to the work of his protégée, recently promoted Special Ops Captain Donald Shepard, the man responsible for bringing down Marcus La'Dek eleven years earlier. As President Vance approached the table, his two advisors stood to their feet in respect.

"Well, how did it go, sir?" Asked General Kirkland as the three of them took their seats.

"I wanted to knock that smug grin off Zek's face," the President said with fists still clenched.

"That well, huh?" Sara asked rhetorically.

"They gave us less than seven months to recover the stolen prototype, or face expulsion," said the President, referring to the galactic cycle he was given by Zek'Ren. He sighed and rubbed his throbbing temples. "So, what's the update, Sarah?"

"The Lyrians believe we staged the theft of the Benton Chamber. All in some elaborate scheme to shift the balance of power within the GPC to our favor," said Sarah while skimming over her data pad.

"Yeah, right," said Vance. "Zek's been looking for an excuse to cross our borders for years. He's just pissed that it was good old-fashioned human ingenuity that solved a problem their greatest minds couldn't."

"Be that as it may," Sarah continued, "they're convinced we plan to dangle the technology over their heads, forcing all member races of the GPC to rely on us to save their dying planets."

"And it gets worse, sir," said General Kirkland as he stood to his feet. He activated the main view screen on the far side of the room.

The screen showed surveillance footage from one of their stealth drones taken just hours before their meeting. President Vance watched as a massive buildup of Lyrian warships gathered near the ISL border. The screen was filled with craft ranging from angular-shaped fighters performing combat maneuvers around massive carriers to enormous dreadnought battle cruisers capable of laying waste to entire fleets of starships.

"The Lyrians say these are nothing more than training exercises. But our reconnaissance teams believe this to be the likely precursor to an incursion of ISL Space," Kirkland concluded.

"Our spies within the Empire concur. We believe they have no intention of honoring the galactic cycle they gave us to find the chamber. I'd say we have four weeks at best," said Sarah.

President Vance couldn't believe what he was seeing and hearing. He began studying the intelligence reports given to him by Sarah.

"So, let me take all of this in for a moment," said Vance. "The Orion's Shield mercs that stole the device are still in the wind. And now their biggest rival, The Dragen Alliance—led by that Lyrian nutcase, Ar'Gallious—is in the hunt as well?" he asked.

His advisors nodded in agreement.

"And on top of all of that, we now have the entire Lyrian Empire salivating at our doorstep?"

"Yes, sir," both Sarah and Kirkland responded in unison.

"Tell me, people. Is there any good news at this point?"

"According to Sarah's high-level snitches close to Orion's Shield, the chamber has yet to be delivered to Teric due to our increased presence in the Outer Core. They're in hiding, sir. And we're very close to finding the safe house," said General Kirkland.

President Vance stood to his feet after entering several commands into the holographic keyboard projected above his section of the table. In the center of the room, a life-sized image of the Benton Chamber appeared in front of the small audience of three. It was a cylindrical chamber about five feet in length, with a digital interface and thick cables on each end. It was a simple design from its outer appearance, but the inner workings of that piece of advanced technology, the brainchild of Doctor Jakob Benton Jr., promised to be the answer to the greatest energy crisis the galaxy had ever known.

"When the Krillium ore supplies dry up in the next twenty years, *this* chamber will prevent the catastrophic failures of literally thousands of Planetary Terraformers throughout the galaxy," said President Vance while gazing proudly upon Dr. Benton's magnificent achievement. "Any word on the whereabouts of Dr. Benton and his research team?"

"Nothing yet, sir," said Kirkland. "They went missing during the fire fight at Maxis Labs."

"We'll find the team and the chamber, sir," said Sarah. "And despite the Lyrians' baseless accusations, we *will* share the technology in accordance with the GPC Scientific Mandate of 3030."

"This thing is more than an answer to a problem," President Vance said as he moved closer to the floating three-dimensional likeness of the Benton Chamber. "It's our one shot at respect within this dammed interstellar community," he concluded. He walked to the table and deactivated the projection. "Dismissed."

After the meeting adjourned, the two senior advisors left the president's office with plates considerably fuller than when they first arrived. As Sarah Miller scurried off to the intelligence wing of the Citadel, Kirkland answered an incoming transmission on his personal comm device. It was Captain Donald Shepard.

"Sir, we need to talk. We may have just caught a break," Donald said, much to the relief of his commanding officer.

They needed results, and they needed them fast. Kirkland knew that Donald wouldn't have reached out if he didn't have a major update. "Excellent, captain. Meet me at the Citadel Officers' Club."

CHAPTER TWENTY-SEVEN

The small office of La'Dek Transports was packed to the seams with farmers from the nearby planet, all looking to have their newly harvested goods shipped to neighboring star systems. While Samantha, Elizabeth and Kayla were hard at work in the rear of the office, Marcus was wrapping up with a farmer at his desk. Without looking up, he greeted the next customer.

"Welcome to La'Dek Transports," said Marcus. "Contents of your cargo?"

"It's a pickup actually," said a strangely familiar voice. "And the contents are a bit, complex."

Marcus looked up to see none other than his former second in command and ex-friend, Daren West. Part of him wanted to toss Daren out of the nearest airlock for getting him busted on Titan eleven years ago. But he was so shocked to see him, that he sat there, completely speechless.

"Nothin' to say? C'mon now, I need a little customer service over here," said Daren, joking as if no time had passed.

"What the hell are you doin' here?" asked Marcus sharply.

"Well now," said Daren. "This isn't exactly how I imagined our reunion."

"Reunion?" asked Marcus, completely livid. It took everything he had to keep his voice down. "This ain't no reunion. The best thing you can do is turn around and walk out that—"

Samantha walked over to Marcus' desk and interrupted the conversation. She looked at the dark-skinned, handsome gentlemen conversing with her brother and smiled. She had a thing for tall, muscular men with bald heads.

"Good morning, and welcome to La'Dek Transports. Are you a friend of Marcus?" she asked.

"You could say that," Daren replied, while trying to suppress the smile cracking his lips. He gently grabbed and kissed her hand. "Daren West. Pleased to make your acquaintance."

"Oh, and a gentleman too," Samantha said to Marcus who looked visibly annoyed. "I'm Samantha Brown-La'Dek, Marcus' sister."

Daren took one look at his former leader and immediately knew this potential quarry was off limits. Marcus swiftly moved in to break up the handholding. He grabbed Daren by the arm, leading him toward the back office.

"Take over for me, Sam," Marcus said, not even waiting for a response.

"It was a pleasure meeting you," Daren said as he was forcefully led away.

Upon entering the office, Marcus slammed the door shut. Before Daren could open his mouth, Marcus caught him with a right cross that nearly separated the mercenary from his boots.

"What's wrong with you?" Daren asked as he found himself on the floor, wiping blood from his lip.

"What's wrong with me?" Marcus replied, appalled at the ridiculous question coming from the man that was supposed to be his brother. "Do I really need to say?"

But Marcus' question was followed only with a long-discomfited silence, as if his words had penetrated some unseen, yet formidable set of armor that Daren had tried in vain to reinforce over the years. The truth was that Daren had spent more than a decade agonizing over his actions on Titan.

"Look, man, *sorry* doesn't begin to—"

"Sorry?" asked Marcus. "*Sorry* doesn't give me back the years I spent rotting in that Martian hellhole."

"C'mon, bro..."

"We ain't brothers," Marcus roared. He took a deep breath to compose himself, then turned his back to Daren. "You know, your father taught us things that probably no child should ever learn," Marcus continued. "But he did teach us something about loyalty. Guess you were asleep during that lesson."

"He died, Marcus," Daren said, "a few years back."

The message knocked the wind out of Marcus. He staggered to the lightly padded chair behind the lone desk in the room and took a seat. After a lifetime of brushes with death, Marcus figured that old pirate to be immortal.

Jax was more than a mentor to him; he was a father. He knew Daren was always jealous of that relationship, so it was no surprise that his so-called *brother* would drag his feet with delivery of the news. "How did it happen? When?"

"Phaedren's disease, 'bout seven years ago," Daren somberly replied, referring to a drug-resistant plague attributed to ancient biological weapons testing in the outer core's Feriss sector. "Must've caught it while runnin' guns on those backwater planets near Feriss."

"After all we've been through, you could've at least sent a message. I think I deserved better than that," Marcus snapped.

"What do you want me to say? I'm tellin' you now," Daren responded in kind. "But I know what you wanna hear," he continued. "You wanna hear about how on his deathbed all Dad could talk about was you, right?" Daren asked while pulling out an XP-90 pistol from his waist holster, slamming it on the desk.

Marcus stared momentarily at the weapon, then stood to his feet. He moved face to face with Daren, ready to take things to the next level if necessary.

"Or do you want to hear how his dying words included letting me know how stupid I was for getting you busted?" Daren continued, never backing down from his former leader. "What do you want from me? Blood?" Daren asked, pointing to the pistol on the desk. "Well, there it is, Marcus."

In one swift motion, Marcus stepped behind Daren and delivered a swift low kick to the back of his legs that brought the disgraced mercenary to his knees. He snatched the pistol from the desk and dragged Daren to the center of the room by his collar, pinning him face down on the floor. Marcus readied the weapon and pressed its barrel hard against the base of Daren's skull, ready to do the one thing he dreamed of during those cold nights he had spent locked away in stasis.

"Ten years, Daren," Marcus said.

"Go ahead, Dek," Daren said in a low, surrendered tone. "Do both of us a favor."

Whatever Daren had been into since Titan, it had obviously taken a tremendous toll on him. Marcus had never seen him so dejected. He sighed and dismantled the weapon, tossing it to the carpeted floor.

"You delivered your message," said Marcus. "Now go."

Daren stood to his feet after retrieving his sidearm. He traveled a great distance for this meeting, and he wasn't about to leave until he'd accomplished his real goal.

"Don't get mad," Daren began. "But I heard about your company's financial woes, and about your niece."

"You spyin' on me now?" Marcus asked, knowing Daren had extensive contacts on Stratus One. "And let me guess, you have the perfect job that'll solve all our problems, right?"

"Just hear me out."

"We don't need your kind of help."

"It's a simple cargo run, completely legit. My usual guy is nowhere to be found, and I—"

"First time I see you in all these years, and you need a favor?"

"It's not like that," said Daren.

"Whatever you're into, you're not dragging me or my family into it," Marcus said, refusing to let Daren speak.

"You know, D," Marcus continued. "After being stuck in stasis for almost three years, I had an epiphany. You know what I realized?"

"What's that, Dek?" Daren said amid a sigh of irritation.

"That I didn't want to spend the rest of my life there," Marcus said, though part of him truly *was* desperate for the money. "I grew up, D. It's time you did the same."

"Well, we all don't have something to go back to, do we?" Daren said forcefully. "The life you had on this station, the life you ran away from as a kid? I would've killed for. But I'm tired, Marcus. I just want out. And this job is my shot."

As Marcus listened to him speak, he wasn't sure why he hadn't thrown Daren out already. He had his sister and niece to protect, so he turned his back to Daren.

"C'mon, Dek, use your head," Daren yelled. "At this rate, your family will be out of business in a year at best."

Marcus hated that Daren was right. Even more, he hated the fact that he was seriously considering taking him up on the offer. Daren knew him well. He was up against a wall and running out of time. The silence spoke volumes. Daren knew he touched a nerve. Though he regretted having to use Marcus' woes to get what he wanted, he had too much riding on this job and wasn't about to leave that office until Marcus said yes.

"Can't believe I'm actually considering this," said Marcus, appalled that he gave voice to those words. "Alright, let's hear it. But I'm warnin' you, Daren."

"You don't even need to say it," said Daren, knowing exactly where Marcus was headed. "Like I said, your end is completely above water. No illicit cargo involved. Here is the payout," Daren said, as he slid Marcus a data pad with the contract offer.

Marcus studied the data file for a moment.

"So, we good?" Daren asked.

"That's an awful lot for something you can carry on your back," Marcus said after studying the cargo's dimensions. He started to question the entire job but changed his mind. "Forget it. We'll take the job."

"Excellent," Daren said, amid a sigh of relief. "Look, I can't give you back those ten years, so think of the money as my way of saying I'm sorry. And after this, I promise you'll never see me again," Daren said as he headed toward the door. "And for what it's worth, thanks."

"Don't thank me," said Marcus. "It's just business, right?"

"Yeah. Just business," Daren said as he prepared to close the door. "I'll finalize the details with your sister out front."

CHAPTER TWENTY-EIGHT

Inside the lavishly decorated Officers' Club within the east wing of the Citadel, Donald waited patiently in the dining area for General Kirkland to arrive. Kept company by the lone cleaning droid across the room, Donald listened to the soft hum of the built-in scrubbers attached to the droid's arms as it silently buffed the chrome countertops of the bar to a pristine shine, as if completely oblivious to the captain's presence. Moments later, the general arrived, moving with a hurried stride.

"Tell me you have something, Shepard," said General Kirkland, completely dispensing with the pleasantries.

"As a matter of fact, we do," Donald replied. "I think we found it, sir."

"The actual safe house?"

"That's correct. It's on a small planet in the Outer Core. If it checks out, I can have a strike team there in less than an hour."

The captain's words could not have come at a better time. For seven weeks they had been trying to track down the Orion's Shield terrorists that made off with the Benton Chamber from Maxis Labs. They had received multiple tips from reliable sources that the mercs were hiding out somewhere near ISL-controlled space, though pinpointing their exact location proved a difficult task. Donald's team exhausted nearly every resource they had to find that safe house. Lucky for them, criminals talk, and for the right price, they would sell their own mothers without the slightest hesitation.

Once again, Kirkland's most trusted soldier came through when he needed him most.

"The president has us under a quantum-scope on this one," said Kirkland.

"Don't worry, sir. We got this."

"Excellent, Captain. What's the name of the planet?"

"Protos Four. I'm on my way to hook up with our teams in that sector now."

After mutual salutes, Donald hurried off to continue his mission to Protos Four, confident the Benton Chamber would soon be back in ISL hands.

CHAPTER TWENTY-NINE

"Protos Four?" Samantha asked after reviewing the flight itinerary submitted by Daren. "Didn't think there was anything out there but farmers and cows. Your friend gettin' into the dairy business or something?"

Marcus was in the rear of the office making coffee, wishing his overly curious sister would stop with the incessant questions. He poured two cups and walked over to Samantha.

"You never know with that guy," Marcus said as he handed over the steaming cup of java.

"I mean, don't get me wrong, the pay for this run is more than generous," said Samantha. "In fact, we could begin expanding our fleet with this. But who pays this kind of money to transport something this small? Doesn't it seem—"

"Leave it alone, Sam," Marcus gently demanded. "We need this."

"I know we do. But how well do you really know this guy?"

"Very well," said Marcus. As he stood in momentary silence, the multitude of misadventures and near-death experiences he and Daren had faced over the years raced through his mind.

He took a long sip from his mug then grabbed his coat from the nearby office. "I'm headed to the docking bay. They should be there by now."

Samantha gave her brother a hug as he headed toward the exit. Her years spent in the military made her a very cautious woman. In her experience, everything came with a price. And if something looked too good to be true, it usually was. But if her brother vouched for Daren and this job, the least she could do was give him the benefit of the doubt.

"Stay safe out there," she said.

"Always do," Marcus replied, trying to ease the worry he could see building in his sister's face. "I'll be back before you know it," Marcus said as he exited the door of La'Dek Transports, headed to what was sure to be an uneasy reunion with the old crew.

#

The docking bay of Stratus One was always busy, especially during the Interstellar League's holiday that marked its founding. Hordes of ISL citizens flooded the Star Port, looking to return home from their deep-space affairs to join in the festivities. Marcus spent nearly ten minutes dodging endless lines of those passengers as they awaited their earthbound transports. Eventually he arrived at docking bay forty-seven, the location of his beloved ship, *The Indicator*.

Following his release from prison, Marcus spent the better part of a month tracking down Jason Crowley, the pilot from his old crew, Raven Squad. And as usual, he found Jason drunk in a bar on some no-name planet in the outer core where he was no doubt blowing his share from the latest score. Jason was ecstatic, having not seen or heard from Marcus in over a decade, so the two drank and hung out until early morning. It was a good time, but bittersweet as well.

By the end of his visit, Marcus broke news that he was out of the mercenary game for good, and that he'd come for his ship. Jason had become especially attached to *The Indicator*, having been its chief steward for years. But he knew the ship ultimately belonged to Marcus, so he surrendered the vessel without a fuss. After Marcus dropped Jason off at the nearest Star Port, the two friends wished each other well and parted ways.

That was the last time Marcus had seen anyone from the old crew. And now, more than a year later, he was moments away from reuniting with the entire team... and he wasn't sure how he felt about that. As Marcus stood gazing at the ship, lost in thought, he heard them approaching from behind.

"And there he is," said Daren to the rest of the team as they emerged from amongst the crowd.

Jason approached first, wearing his signature black-and-gray armor-plated flight jacket and red knit skullcap.

The articles of clothing were old to be sure, but he said they brought him luck, so he seldom left home without them. The light-skinned man's hazel eyes widened with excitement upon seeing his old leader. He was happy the whole team was together, even if it was perhaps for the last time.

"What's up, Dek?" Jason said as he quickly moved in, trying in vain to lock in a signature chokehold move on his former leader, a feat he often attempted in the old days. But he never could overpower Marcus, and still couldn't. He quickly found himself flat on his back after Marcus sidestepped the maneuver and struck him with an impeccably timed leg sweep.

Jason laughed. "Still got it I see."

"I don't even know why you try, Jay," Marcus said as he helped the embarrassed mercenary back to his feet.

Tony, the Gorean Cyborg, approached next, shoving Jason aside like a rag doll. With blinding speed, the six-foot-seven mass of muscle and living metal moved face to face with Marcus. From a distance, Goreans could almost pass for a large human, but up close, one could see the notable differences. As Marcus studied the sharp, angular features of Tony's face covered by smooth pale skin, the faint golden glow of Tony's solid-colored eyes peered back, as if looking for the slightest sign of weakness.

Then without warning, the cyborg cracked a rare smile as he embraced Marcus with a crushing bear hug, lifting him several inches from the ground.

"Great seein' you too, Tony," Marcus said as he struggled to utter the words under the substantial pressure applied by the cyborg's muscular arms.

"Sorry about that, man," Tony said as he released his grip. "About that night on Titan..."

"Forget about it," Marcus replied. The last thing he needed was to dredge up old memories of the Titan job. "You did what I ordered. We're good."

Tony nodded. Even though he failed to articulate his regret about that night, he was relieved to know that, after all those years, Marcus didn't hold it against them. "Thanks, man," said Tony, feeling as if a massive weight was finally lifted from his broad shoulders. He joined Jason at *The Indicator*'s aft loading ramp, after which the two men entered the ship.

Marcus turned to see Skyela approaching. With smooth brown skin, she was of a darker complexion than that of her half-brother, Jason. Her usually curly black hair was tightly braided in cornrows from front to back, converging into a shoulder-length ponytail. Being an avid runner, she had a lean, muscular build, clothed in brown leather armor reinforced with a nanoplate covering, strong enough to withstand nearly any type of small arms fire.

As she moved in to greet him, Marcus was hit with a flood of emotions that he wasn't expecting, nor prepared to handle. For a moment, he almost lost sight of his purpose for being there. He snapped out of it. He wouldn't allow himself to entertain thoughts that might suck him back into the old life. For once he needed to think about more than himself, so he preempted Skye's advance by addressing Daren.

"We need to get goin'," said Marcus.

Skye was furious after being completely brushed off. She started to make a scene, but Daren stepped in before she could say a word.

"Before we go, let me introduce you to the new guy, James Scott," said Daren.

"Call me Jack," said the merc as he extended his hand toward Marcus.

But the gesture wasn't reciprocated.

Marcus examined the muscle-bound, battle-tested mercenary with the facial scars to match. He was clothed in black tactical body armor, with a duffel bag strapped to his back, containing what Marcus was sure to be weapons. In fact, they were all curiously dressed for what was supposed to be "a simple cargo run" as Daren had so eloquently described it the other day. Despite his gut feeling telling him otherwise, Marcus opted to give them the benefit of the doubt. *I'm sure they're here for security to make the client feel better*, Marcus reasoned in his mind. But still, something didn't feel right about the new guy. He was an unknown, and Marcus didn't care for the unknown, especially aboard his ship.

"Where's Max?" Marcus asked.

"That, my friend, is a very long story. I promise to tell you about it later. But like you said, we need to go," said Daren, successfully dodging the one question he had no desire to answer.

Marcus started to press the issue but changed his mind. At that point, he had one focus, *money*. His family was counting on him, so he was perfectly willing to overlook a few *discrepancies* along the way. After everyone was safely aboard, Marcus strapped himself into the captain's chair on the bridge, followed by Daren who settled into the copilot's seat. After initiating the ship's departure sequence, Marcus activated the communication system.

"Victor-thirty-seven, this is Transport Ship Alpha-two-eight-niner, requesting permission to depart," he said to the station's space traffic control tower.

"Permission granted, Alpha-two-eight-nine. You are cleared for departure," the traffic controller's voice crackled over the radio. Marcus engaged thrusters and slowly piloted the ship through the enormous hangar bay doors. After moving the ship a safe distance from Stratus One, he engaged the hyper-jump engines. And in a brilliant flash of light, *The Indicator* disappeared as it made the jump to hyperspace.

CHAPTER THIRTY

Daren sat quietly in the copilot's chair, staring blankly at the multicolored lights dancing on the instrument panel in front of him. The silence was driving him crazy. Two hours into the trip and the two men hadn't said a word to each other. He couldn't take it anymore, so he decided to break the silence.

"The legendary Marcus La'Dek and his famed *Indicator* gunship," Daren said as he looked around the bridge. "I know you modified her for cargo transport, but she sure was impressive in her day."

But there was no response. Marcus continued to run maintenance routines on the secondary navigational computer, completely ignoring his former business partner.

"He won't admit it, but Jay nearly slipped into a state of depression when you came back for your ship," Daren continued, chuckling as he remembered the look on Jason's face when he broke the news to him that day. "And you know I was pissed, right? We had a job lined up the next week. I mean, I had to practically beg Marco to borrow his busted rig just to do the contract," Daren continued to his inattentive audience of one. "Remember when that guy abandoned us on Korack Three? So, you know he was the *last* person I—"

"Look," said Marcus, interrupting Daren's lone stroll down memory lane. "This ain't that kind of trip. Our ETA is three hours," he said, not even turning to look at Daren.

The message was received loud and clear.

"Okay then," Daren said as he stood to his feet. "I'll be in the cargo bay if you need me," he said as he headed toward the exit.

As Daren departed the bridge, Skye stepped in just before the automated door closed. Marcus continued to work on the computer, completely unaware of Skye's presence. She looked at him for a moment, then spun his chair around to face her. Before he could say a word, Skye delivered to him a passionate kiss eleven years in the making. Marcus was completely taken by surprise, which didn't happen very often.

Part of him wanted to push her away, but he was completely powerless in that moment, so he went with it. The crew had always suspected there was something going on between the two of them, despite their best efforts to keep it quiet. So, in the back of his mind, he prayed no one else would walk through that door, especially Jason. As the kiss intensified, it was suddenly over as quickly as it began. Skye pulled back and caught Marcus with a right hook to the jaw that nearly knocked him from his chair.

"Okay," Marcus said as he examined his jaw to make sure it wasn't dislocated. "I deserved that."

"I haven't even started yet. I like how you just walked past me back there, without so much as a 'hello,'" Skye said. She then proceeded to curse him out at the top of her lungs.

"Can you talk any louder? I'm sure there's someone near Saturn that hasn't heard you yet," Marcus said, trying desperately to contain the situation. But she continued.

"What happened to us, Marcus?" she asked, finally lowering her voice. "*You* said we were in this thing 'til the end," she reminded him.

"C'mon, Skye. You know what happened," Marcus said. "We agreed. One more job and that was it. How was I supposed to know the whole thing would go sideways?"

"You can save that for someone that doesn't know," Skye countered sharply. "We had enough to walk away *before* the Titan job, just you and me."

"Okay, you want the truth?" asked Marcus, finally ready to come clean to the only woman he had truly ever loved. "I was young and stupid," he admitted. "I wasn't ready to give it up. I just wanted more. You know, for us."

"For us? Or for you?" asked Skye, looking deep into his eyes. "We had each other, Marcus. That should've been more than enough."

They stood for a moment in icy silence.

"How I felt about you never faded, by the way," Skye said, finally breaking the silence. "I even sent messages while you were on lockdown."

"I got your messages."

"Wow. And not one response," said Skye. "And here I was thinking they never made it through," she said, as she slowly backed away toward the door. "Didn't think people could just turn their feelings off like that."

As Skye opened the door, Marcus grabbed her by the hand.

"They're not turned off, Skyela," he said, trying desperately to find the words. "I just, have a lot goin' on right now," Marcus said. He winced as the words parted his lips. He wanted to kick himself for saying that out loud.

Skye recoiled her hand, at a complete loss for words. There was so much more she longed to say, but she held her peace. She simply gave him a soul-piercing look, turned around, and walked out the door.

CHAPTER THIRTY-ONE

Two hours later, Daren, Tony, Skye and Jack retreated to a dark corner of the cargo bay, studying holographic blueprints of a large compound on the surface of Protos Four.

"Should be a cakewalk. In and out," said Daren to the team.

"What's up with the pilot?" asked Jack. "I know he's supposed to be your boy and all, but the guy hasn't left the bridge since we left Stratus. You sure he's even down?"

"We don't need him," said Skye to everyone's surprise.

"Let me worry about Marcus. Just make sure you stick to the script. I heard about you," said Daren to Jack.

"You got a problem with me being here? Take it up with, Kull. But from what I hear, this is it for you," Jack fired back.

"The only reason you're here is because Max is down," snarled Daren.

"Oh yeah, coma, right? That's what happens when you work with amateur lab rats that got no business in the field," Jack said, now standing to his feet.

Daren arose from his seat and moved face to face with the brutish mercenary, as if ready to put him through the floor. But Skye moved in to separate the two would-be combatants.

"Stop it, you two," yelled Skye. "All of our necks are on the choppin' block."

"Speak for yourself. I ain't the one that screwed up the first job," said Jack.

"Tell me you aren't as stupid as you sound," said Tony. "You really think Teric is gonna let you walk if we come back empty handed?"

But Jack dismissed the cyborg's words and pulled an XR-17 assault rifle from the duffel bag he was carrying. As he began to perform routine maintenance on the weapon, Skye stepped forward.

"There isn't a soul aboard this ship that wouldn't rather be someplace else," said Skye. "Like it or not, we're stuck with each other, so you may as well strap in and enjoy the ride."

"Just make sure you don't drop the ball out there," said Daren, never taking his eye off Jack.

"Worry about yourself. Once this is over, we'll have our time," Jack replied.

Before Daren could respond, they were interrupted from a voice coming from behind.

"Once what is over?" asked Marcus as he emerged from the shadows of the cargo containers near the team.

Everyone quickly tried to act normal, but it was too late. Daren glared at Jack, then focused on Marcus.

"So, what's goin' on here?" asked Marcus, motioning toward the rifle that Jack was still holding. "Daren, wanna fill me in? And please, don't tell me, *it ain't what it looks like.*"

"C'mon, Dek. Before you do anything crazy..." Daren began, just before hearing the distinctive hum of a round being readied in an XR-17. He turned to see Jack aiming his rifle at Marcus.

"I already see where this is headed," said Jack, with eyes fixed on Marcus. He took a few steps back and addressed Daren. "We don't have time for this. Let's drop him and take the ship."

"Jack, what are you doin'?" yelled Daren. He knew Jack was crazy but didn't think he would go as far as mutiny.

"Sure you wanna do that?" asked Marcus in a low, even tone.

"Very sure. Now tell me, Captain. You gonna play ball? Or do I have to sit you on the bench?" Jack replied.

The rest of the team kept their distance, knowing Jack had a reputation for being unstable. They knew he wouldn't hesitate gunning Marcus down at the slightest hint of danger.

Unfortunately, they needed Jack as much as they needed Marcus, so they opted not to risk an all-out bloodbath.

"Ship's access codes," demanded Jack, knowing the ship was probably on lockdown by now.

If this were five years ago, Jack would have been dead already. But Marcus knew that even though he was still in remarkable shape, he'd been out of the game a long time. His reflexes weren't as sharp as they once were. So, he relied on his experience as he attempted to goad the younger mercenary into making a mistake.

"Can't help you there, boss," Marcus replied.

"I'm sorry, perhaps you thought that was a request," Jack said. Furious at Marcus' defiance, he foolishly moved to within arm's reach. "I said, give me the—"

Before Jack could finish, Marcus shoved the rifle aside while simultaneously stepping forward and hooking his arms beneath Jack's arms. And with a powerful pivoting motion, he slammed Jack to the steel deck with a perfectly executed hip-throw. Jack landed on the ground with such force, the wind was knocked out of him. Marcus grabbed the rifle and shoved the barrel hard into the base of Jack's skull while the shocked mercenary was still on his knees, gasping for air.

"Come again?" asked Marcus as he placed his finger on the trigger.

Tony could see that look in his eyes. He knew Marcus was mere seconds away from decorating *The Indicator*'s deck with Jack's brains, so he reluctantly pulled his pistol on his former leader, hoping he wouldn't have to drop him. Skye watched in disbelief as the entire situation took a sudden turn for the surreal. She placed her hand on her sidearm but didn't pull it out. In unison, they pleaded with Marcus to drop the weapon.

"C'mon, Dek. We got a lot ridin' on this," shouted Tony.

"Let him go, Marcus. I know this ain't Max, but we need him," said Skye as her grasp tightened around her pistol's grip.

By now, Marcus had forced Jack to his feet and was using him as a human shield. Skye's words caused Marcus to hesitate, but he held firm.

"So, that's how it is? We pullin' guns on each other now?" asked Marcus, surprised that is former team would go to such extremes. "Somebody better start talkin', or we can do this right here and now."

Daren looked on as his entire operation was suddenly hanging in the balance, so he stepped between Marcus and the crew.

"Everybody calm down," said Daren. He turned to Tony who appeared to be only moments away from opening fire. "Lower your weapon, everything's cool."

"Cool?" said Marcus. "Did that really just come out of your mouth?"

Tony and Skye completely ignored Daren's command.

"I said drop it on the deck. Now," yelled Daren to Tony and Skye.

They hesitated but complied, placing their pistols on the floor and stepping back.

"Place all of your weapons and gear in that secure container over there," Marcus ordered Tony.

Tony gathered their gear, placed it in the container, and shut the lid. The automatic locks engaged.

"Now, everybody take a seat 'til I get back," Marcus said after shoving Jack toward Skye and Tony. He then turned to Daren. "And you? Step into my office," Marcus yelled, motioning Daren toward the bridge.

Daren headed in first, followed by Marcus. The blast-proof door then closed and locked behind them. Skye examined Jack to make sure he was still useful for the mission, assuming there'd still be a mission. At that moment, they heard the activation of a waste disposal unit, as Jason exited the aft latrine door.

"Whew, y'all may wanna steer clear of there for a little while. Should've laid off that last..." Jason started, but his words trailed off as he noticed Daren missing, Skye tending to Jack, and Tony angrily pacing the floor. "Okay, what'd I miss?"

#

On the bridge, Marcus was completely livid. He engaged the rifle's safety switch and tossed it into the corner. He then dropped the ship out of hyperspace, bringing it to a complete stop less than one astronomical unit from Protos Four.

"I'm all ears, Daren."

"I was gonna tell you, I swear."

"I knew I couldn't trust you," Marcus yelled. *"All money ain't good money.* Ain't that what they say, D?"

"We're in a bad way, Marcus," replied Daren, realizing there would be no talking his way out of this one. "We're in deep with Teric," he continued, nervously pacing the floor of the spacious bridge. "And of course, his psychopathic right-hand man, Samson Kull, vowed to, and I quote, 'adorn the halls of Teric's Grand Chamber with our entrails' if we don't deliver this time."

"This time? What the hell are y'all mixed up in?" asked Marcus, seeing the uncharacteristic look of genuine fear in Daren's eyes.

"Alright," said Daren, deciding to finally come clean. "You heard about that missin' prototype, The Benton Chamber?"

"You mean that tech that's been all over the news?" Marcus asked, bracing himself for what was coming next.

And with Daren's shameful nod, his worst fears were realized, knowing full well that Daren had something to do with its theft. Marcus staggered a few steps back, bracing himself against the navigation console as reality hit, *they could all end up in prison for the rest of their lives*, and that's if the ISL gets to them before Orion's Shield does.

"What's wrong with you, Daren?" Marcus yelled as he flopped down in the captain's chair. "They're talkin' about going to war over that thing."

Daren took a seat beside him in the co-pilot's chair. "I know. I wasn't thinkin'. I saw it as..."

"As what?"

"An opportunity," Daren replied. "You know, to step out of the shadows; to strike out on my own, like we used to talk about back in the—"

"That was a long time ago, D," yelled Marcus. "We ain't kids anymore."

They regarded each other uncomfortably for a moment, then Marcus leaned back in his chair as if he had all the time in the world. He was determined to know how it all went down, including what happened to Max.

"From the beginning, Daren," Marcus said with arms folded.

Daren stood to his feet, knowing it was time to face the music. He took a deep breath and proceeded to recount everything that led them to that point.

CHAPTER THIRTY-TWO

Seven weeks earlier...

Deep in the recesses of the outer core on a planet known only by its stellar designation, Beta-Three-Four-seven-nine, a large castle-like structure carved into the side of a mountain sat silently in the distance. It was the headquarters of Orion's Shield. Waterfalls from the surrounding mountains cascaded between the superstructure's highest towers located on the structure's outskirts. The zenith of those towers peaked barely above the mist that filled the area both day and night.

The shielded outer perimeter wall featured towering platforms at five-mile intervals, each armed with gargantuan pulse cannons capable of shooting down any capital ship that may be lurking in space near the planet. But the base's chief defense was its state-of-the-art sensor-scrambling fields capable of fooling the most sophisticated of scanners into thinking the area was as devoid of life as the rest of the planet. It was a simple ruse, but it kept the Orion's Shield base of operations from being encroached upon for centuries.

Inside a vehicle hangar made of metal and stone, located just inside the northwest perimeter wall, Daren and his crew—Tony, Skye, Jason and Max—sat chatting amongst themselves. They awaited a briefing from Samson Kull, the man whose power and authority was second only to Teric Winters.

"Sure is taking his time gettin' here," said Jason as he anxiously paced the stone floors of the empty vehicle bay. "I thought this was supposed to be some big-time op—"

"Would you sit down and relax?" said Skye. "He'll be here when he gets here."

"Patience, Jay," said Daren, displaying an aberrant calmness about the upcoming mission. "Don't forget, we beat out a lot of crews to get here," he said while motioning Jason toward the empty chair to his left. "If this goes right, we'll be set for life. Hell, it's not like you have anything better to do."

Jason couldn't argue with that, so he moved toward his chair, but before he could sit, Samson Kull entered the room, flanked by two of his most trusted personal guards.

"On your feet," shouted one of the guards, prompting everyone in the room to stand up at attention.

Samson was a former ISG supreme commander and ran the forces of Orion's Shield like a well-organized military. Everyone, especially the mercenaries, hated the structure, but it was the cost of doing business with the well-paying organization. Samson slowly limped to the center of the room. He looked to be a man in his early sixties, a veteran of many battles, though many believed him to be hundreds of years older, just like Teric Winters. It was said they both underwent extensive medical procedures to unnaturally prolong their lives. But no one knew for sure, nor had the courage to ask.

"As you were," said Samson, prompting Daren and his team to return to their seats. "What I'm about to tell you doesn't leave this room," Samson said in his booming, yet raspy voice.

One of the guards activated a holographic presentation.

"Your target is in the Kazar system, home of the top-secret private research facility, Maxis Labs," continued Samson, pointing to the holographic layout of a wide-spread research compound.

The building's image was then replaced with a cylindrical-shaped device. "You are after critical technology that could very well turn the tide in our struggle with the Interstellar League," said Samson. "It's called the Benton Chamber."

"Doesn't look like much. What does it do?" blurted Jason, much to the dismay of the rest of the team.

"Its function, Mr. Crowley, is none of your concern. Just know that Mr. Winters desires this above all else," said Samson as he slowly made his way in front of the team. "We had a deal with Maxis to acquire the tech before he was to turn it over to the ISL. But that arrangement fell apart when President Vance learned of his treachery."

"How do we know the government hasn't already raided the lab?" asked Skye.

"Because the only thing holding them back is the formidable army of Maxis Corp lawyers keeping the government tied up in court, arguing over ownership of the chamber," said Samson. He motioned toward one of his guards to deactivate the presentation. "But it's only a matter of time before this little *distraction* is resolved."

"Which gives us a small window to get in and out before the ISG storms the place," said Skye.

"Very perceptive, Ms. Evans," said Samson. "As far as Maxis knows, our dealings ended when he was discovered. And while they aren't expecting us, you can rest assured his security forces will defend the technology 'til their dying breath."

"Just the way we like it," Daren said as he stood to his feet. "When do we leave?"

"I appreciate your enthusiasm, Mr. West. But first, let me introduce you to your pilot for this mission."

"With all due respect, we already have a pilot that's more than—"

"Mr. Thomas Lance," Samson continued as if Daren had never spoken.

Tommy walked into the room, clad in silver and black. As always, he wore his signature shades that provided a virtual heads-up display, feeding him tactical information about his surroundings. He approached the team, brushing a section of his long, blond hair to the side. He was accompanied by his petite copilot, Lana Doran.

Lana had short, crimson-colored hair that her friends said matched her fiery personality. Her perfectly tanned yet unblemished skin made her a sight to behold. She was dressed in a dark gray form-fitting jumpsuit that accentuated the flawless curves of her toned body.

"I'll leave you to it, love," said Lana in her soothing Australian accent. Before heading toward the ship, Lana gave Tommy an inappropriately timed passion-filled kiss that triggered a few disgusted looks from Daren and his crew.

Lana casually departed the room with Jason's eyes fixated on her every sexy, yet graceful, stride. He stared so hard, Skye had to discreetly elbow him just to get him to look away.

Tommy approached Daren and the others just as Lana exited the hangar.

"Well hello again, Raven Squad. Or are you still goin' by that these days?"

No one dignified Tommy's question with a response. Jason turned and nudged Tony.

"Isn't that the guy that trains the snipers around here? What the hell does he know about piloting a starship?" whispered Jason.

Thanks to Tommy's electronically augmented hearing, he caught every single word.

"No offense, kid. But I was piloting starships when you were still knockin' over food stands on Earth."

"Why don't you leave the flyin' to the pros?" said Jason. "Maybe we'll call you when someone orders a political assassination or somethin'."

Tommy ignored him and instead, approached his sister.

"You've gotta be kiddin' me," said Skye to Daren, recognizing Tommy from the bar the other night.

"Offer stands, you know," said Tommy with a slight grin on his lightly bearded face.

"Like I said, never gonna happen," replied Skye, returning a halfhearted smile of her own.

Tommy continued to grin while examining her exquisite figure, when Jason stepped between the two of them.

"Wanna put your eyes back in your head?" growled Jason.

"At ease, kid," said Tommy. "Just playin' around."

Daren was irritated. He was perhaps the most familiar with Tommy than anyone in the room. And he had every reason for wanting Tommy as far away from the job as possible.

"Mr. Kull," pleaded Daren, "I assure you—"

"You leave in two hours," said Samson. "Your mission is the Benton Chamber and the research team. Details of your targets, including the infiltration plan, are being uploaded to your data modules as we speak."

Samson exited the room along with his two guards, leaving Daren to prep his team for the mission.

"Well then," said Daren following a sigh of disappointment. He then approached the crew. "Shall we?"

#

Twelve hours later, Daren and his team were fast approaching the fourth planet of the Kazar system, deep in Outer Core space, far from the watchful eye of the ISL. Within thirty minutes they were on the ground about five kilometers outside Maxis Labs.

The team filed out of the port hatch of the gunship that was disguised to look like a radiation containment vessel. Tommy exited last and approached Daren.

"We'll rendezvous here in two hours. Be here, or we're gone," said Tommy.

"Don't worry about us, just make sure you keep the thrusters burnin', *Specter One.*"

"Better get a move on," said Tommy. "Clock's tickin'."

Daren briefly looked at the arrogant mercenary standing before him. He wanted nothing more than to smash those stupid shades he always wore while they were still strapped to his face.

The very notion of working with Tommy again made Daren's skin crawl. In fact, he and Tommy—better known by his call sign, Specter One—were the ones responsible for the failed job on Titan that resulted in Marcus being locked up, a truth he never disclosed to the rest of the team.

For years, he cursed the day he let Tommy talk him into running that side job on Titan, all to impress the highly secretive criminal organization, *The Company.*

They never did gain entrance into The Company, and the two former business associates had a falling out afterward.

Daren continued to glare at Tommy as he entered the ship. He snapped out of it and turned his attention back to the task at hand.

"Alright, Ravens," Daren said to the team, "let's do this."

#

A short time later, the team closed in on the one section of the outer wall that was, per Skye's intel, a virtual surveillance blind spot. Maxis Labs itself was a series of interconnected high-rise buildings constructed of ballistic glass and metal. One could get lost for hours without firsthand knowledge of the facility. But fortunately for the team, the one piece of reliable intelligence they did have was the exact location of the Benton Chamber and the research team, all of whom were being held against their will, waiting to be sold to the highest bidder along with the prototype. In this case their targets were underground in an expansive subterranean laboratory, the place where the real research was conducted.

Tony reached into his equipment pack, pulling a large rectangular-shaped object that he magnetically affixed to the wall. He was unsure if the experimental breaching device would even work, but Max assured him that her design was flawless. He pressed the lone button at the device's center, causing five-foot arms to extend from the ends of the apparatus, each containing holes at one-inch intervals aimed toward the wall. And after calculating the wall's thickness, the breaching device fired multiple beams of intense light from its arms into the wall. The beams superheated a circular section of the wall, causing it to liquefy as the device rotated itself counterclockwise, resulting in a perfectly bored hole through which the team had little trouble traversing.

On the other side of the wall, the team found themselves inside a fully stocked motor pool filled with mostly wheeled and hover vehicles. Skye led the team to the far side of the motor pool to a massive metal door leading to the maintenance section of Maxis Labs. It was time for Skye to work her magic. She had determined beforehand that wireless scything of the doors was out of the question. For their entrance to appear routine, she needed a hard-wired connection. Once inside, their forged electronic badges would allow them access to the lower floors.

Skye pulled out her trusted portable scything device and attached it to the door's control panel using a patch cable with a universal connector. Within seconds, she cracked the door's code. One by one the team filed in with their rifles at the ready. After making sure the maintenance section was clear, Daren addressed the team.

"Tony, you're with me. We're goin' after the Benton Chamber. Skye, you take Max and Jay to secure the research team. We'll meet back here in thirty," said Daren.

The team acknowledged their orders and proceeded down the turbo elevator to the lower facility where they split up to go after their targets.

#

An hour later, Tommy landed the ship at the rendezvous point. He checked the time on the bridge console and turned to his copilot, Lana.

"They're cuttin' it close," he said.

He removed his tactical shades, revealing eyes that were pale gray, with jet-black pupils. While most women were mesmerized by the exotic look of his eyes, the discoloration was due to the staggering amount of cybernetics he had installed behind his irises, extending back to his optic nerves. The advanced ocular implants were designed to interact with his shades via circular interfaces located near the base of his skull.

The bands of his shades, which wrapped around his head, attached to the interfaces, providing him with many advantages, including enhanced eyesight and information acquisition, seen only on the level of military-grade Android Combat Units. But Tommy's improved visual acuity came at great cost. He was, in a sense, a prisoner to his shades, as prolonged time with and apart from the eyewear at times resulted in debilitating headaches.

"I'm sure they'll be here," Lana said as she moved toward him in a provocative manner, lost in his entrancing eyes.

And despite the painful pressure slowly building behind those eyes, Tommy knew what Lana wanted, and he was more than happy to oblige.

#

In the subterranean lab, a lone security guard feverishly sprinted down the main hallway toward a red box affixed to the wall. His heart pounded as he flinched at the sounds of yelling and weapons fire coming from around the corner. The out-of-breath guard opened the box and pulled the switch, causing alarms to go off all over the building.

Seconds later, Daren and Tony rounded the corner toward the guy who had sounded the alarm. Daren had a large black duffel bag strapped to his back containing the Benton Chamber, and an XR-17 assault rifle in his hands. Tony was armed with his trusty heavy repeater. The over-matched guard pulled his sidearm and attempted to fire, but Daren and Tony gunned him down before he could get off a single shot.

165

"Skye," Daren yelled into his comm device. "The facility's goin' on lockdown, I need you to reactivate the lift."

Seconds later, the panel lights on the turbo-lift controls turned green and the doors opened.

"You should be good," Skye shouted back over the comm device. "Get topside, we'll be right behind you," she yelled amid the chaos of the firefight in which she and her team were engaged.

Daren and Tony took the turbo-lift to the top floor. Moments later, Skye and her team exited the adjacent lift, but the research team wasn't with them. There was no time for Daren to inquire about the failed objective. The only thing on everyone's mind in that moment was getting out of that lab in one piece. Daren pulled out his comm device.

"Specter One, we need emergency evac," Daren said, but there was no response. They were being jammed.

The team sprinted through the motor pool, followed by a hornet's nest of guards, some of which began mounting military-style vehicles in the area.

The team ran toward the hole they had made when they first entered the facility grounds. Tony and Max stopped to return fire, covering the team's escape. While Tony laid down heavy fire with his massive rifle, Max launched thruster-propelled grenades from her weapon toward their attackers, taking out several guards in the process. The sounds of the explosions were deafening. Seconds later, a hover-jeep emerged from the thick black smoke caused by Max's grenades. The gunner manning the tactical missile launcher at the rear of the jeep opened fire. Tony saw the missile bearing down on Max's position and yelled as loud as he could.

"Max, get out of there," Tony roared. But he was too late.

Max tried to run, but the round exploded midair near her left side. The concussive force of the blast flung Max several feet into the air, sending her careening to the ground like a ragdoll. Tony switched to anti-armor rounds and lit the vehicle up, causing it to explode in a massive fireball. After reaching the hole, Daren turned upon hearing the explosion. His heart sank as he saw Max crashing to the concrete floor. He shoved the duffel bag containing the Benton Chamber into Jason's chest.

"Take it," Daren yelled. "You and Skye get to the other side and try to reach Tommy."

Without waiting for Jason or Skye's response, Daren darted toward Max.

Within seconds, he reached his fallen comrade, only to find her a bloody mess. The left side of her face had sustained heavy damage, rendering her almost unrecognizable.

He quickly examined her further and found that her left arm was completely blown off from the shoulder down. The very sight of Max brought tears to the hardened mercenary's eyes.

For Daren was especially close to Max, having personally recruited and trained her himself. He struggled to hold it together.

"No, no, no," said Daren to himself. "Not like this," he repeated as he checked her vitals.

To his right, he saw Tony raining fire down upon their attackers. The cyborg fought like a man possessed as he continued to thin out the herd of guards as they spilled out of the vehicle bay from the underground laboratory.

To Daren's relief, Max still had a pulse, though it was very weak. He hoisted his crewmate upon his shoulders and headed through the hole in the wall. Tony continued to cover their escape, then rushed to join them just before reinforcements swarmed the area. Both Skye and Jason were devastated upon seeing the condition of Max. But there was no time to lament her condition.

"No luck reachin' Tommy," Jason's voice cracked as he continued to stare at Max's limp body draped helplessly across Daren's shoulders. "We need to clear the jamming signal's range."

Daren acknowledged, and the team sprinted toward the rendezvous point. Within minutes, they found themselves overwhelmed by more guards, slowing their escape considerably.

Daren checked his comm device. He had a clear signal, so he hailed Tommy.

#

The emergency alert blared throughout the bridge, startling Tommy and Lana. Tommy shoved the woman away and answered the hail.

"Go for Specter One."

"Tommy," Daren shouted over the radio. "Get out here now. We're pinned down."

"On my way," replied Tommy, as he quickly retrieved and strapped on his tactical shades from atop the nav-computer.

Moments later, they were airborne. Tommy ordered Lana to engage maximum thrusters. And with a violent burst of speed, the gunship thundered toward Raven Squad.

#

Near Maxis Labs, Daren and his team retreated to a nearby abandoned structure that offered plenty of cover. From that location, they fought off the incoming guards.

While the rest of the team engaged the oncoming security forces, Skye's combat medic skills were pressed to the limit. She worked frantically on Max to keep her from bleeding out. Tommy arrived moments later with twin energy-based cannons fully deployed, along with a host of armed missile pods. He maneuvered the ship above the team, facing the oncoming horde. Tommy hit the security forces with everything he had, while strafing left and right. The energy cannons ripped through the guards, while the missile pods made short work of the light vehicles.

The coast was clear, but Tommy knew they'd be back with the big guns soon enough. He set the ship down to pick up the crew. After the last Raven Squad member was aboard, Tommy retracted the loading ramp and shut the doors. Moments later, he piloted the ship toward the stars, where he planned to make a quick hyper-jump as soon as they cleared the planet's atmosphere.

#

Inside the passenger bay of the gunship, things were hectic as Max started to go into shock.

"Hold her down," Skye yelled to Tony as she frantically tried to stabilize their friend.

Max was losing blood fast, leaving them no time to administer a sedative and to wait for it to kick in.

"She's losin' too much blood. Somebody grab my medical laser," yelled Skye.

Jason rummaged through Skye's equipment pack until he found the small oblong-shaped device. He tossed the med-laser to Skye.

"Sorry, girl. This is gonna hurt."

Skye used the laser to stop the bleeding, causing Max such agony, that she let out a gut-wrenching scream just before passing out. Skye administered a sedative to keep Max unconscious, allowing her to prep the arm for a crude field dressing.

After making the jump to hyperspace, Tommy turned control of the ship over to Lana and joined the crew in the passenger bay.

"What the hell happened back there?" yelled Tommy.

"Some friggin' security guard spotted us on our way out," replied Tony.

"Damn, nobody was supposed to be down there," said Tommy.

"Yeah? Well I guess they didn't get the holo-trans," said Skye as she continued to work on Max. "We had the scientists but got separated during the firefight. They made a run for it."

"We have to turn this ship around," yelled Daren to Tommy. Daren's heart felt as if it were lodged into his throat. He watched in horror as Skye did the best she could, given the limited medical supplies they had on hand. "I know a guy with a fully stocked med bay—"

"No can do, my man," said Tommy. "When the ISL catches wind of this, the entire sector will be crawling with Fleet ships."

"Are you crazy?" yelled Daren. "If we don't get her help, she's—"

"If we don't get to that meeting spot, we're all dead," said Tommy. "You should have left her," he continued, looking toward Max. "We barely made it out with you luggin' that dead weight around."

Enraged, Daren tackled Tommy to the floor.

Jason rushed in to pull Daren away before he could beat their pilot to death, which under different circumstances, he would have happily stood by and watched. Tommy picked himself up from the floor, wiping the streaming blood from above his eye.

"C'mon, man. Are you serious?" said Tommy, once again pointing to Max on the deck. "She knew the risks when she signed up," he yelled to their dejected team leader. "We all did."

Daren stumbled back to the bulkhead and slid into a sitting position as he watched the entire mission spin violently out of control.

"We lost the scientists, but we still have the chamber," said Tommy, motioning toward the black duffel bag in the corner. "You know it's our only play."

Daren looked at Tommy for a moment then focused on his dear friend and protégée lying helpless on the deck.

At that moment, he knew what he had to do.

CHAPTER THIRTY-THREE

Present day...

As he came clean to Marcus about everything, Daren was careful to omit the parts about him and Tommy stabbing Marcus in the back eleven years earlier. The two men sat in silence on the bridge of *The Indicator* as Marcus took it all in.

"And after you left the planet?" asked Marcus.

"Skye stabilized Maxlyn. And against my better judgment, we proceeded to the outer core drop point."

"So, what happened?"

"Tommy Lance is what happened," yelled Daren, reflecting upon the monumental betrayal he and his team suffered at the hands of that traitor. "So as not to bore you with the details, the guy had his two Lyrian buddies ambush us at the site, just before Samson and his crew showed up for the exchange. They made off with the Benton Chamber."

"Let me get this straight, you got jacked in the O.C. by a pretty-boy and a couple of Lyrians?" asked Marcus, completely sure that Daren had finally lost his edge.

"Not just any Lyrians. We're talkin' Dragen Alliance here," Daren fired back, completely shutting Marcus down.

The Dragen Alliance, a splinter criminal faction of the Lyrian Empire, led by the legendary warlord, Ar'Gallious, was bad news by anyone's standards. Not to mention, they were Orion Shield's biggest rival, as the two factions constantly battled for control of the Outer Core Star Systems.

The whole story went from bad to worse as Marcus began to truly fathom the level of trouble in which Daren and his team had found themselves.

"The only reason Teric didn't string us up from the rafters was because I slipped the Benton Chamber into a portable Mk17 storage unit, before it was taken. The security software can't be scythed, and only I have the access codes."

"And let me guess, you traced the container to Protos Four?" asked Marcus as he put the pieces together in his mind.

"Exactly," said Daren. "They've been layin' low, trying to avoid the ISF patrols."

"And that's what this little trip is all about. Gettin' back at the guy that got one over on you."

Daren began to speak, but Marcus cut him off.

"And when exactly was I supposed to find out? Before or after the shooting began?"

"If I told you up front, you wouldn't have helped."

"You're damn right I wouldn't have helped," yelled Marcus. "But that was my choice to make, Daren. Not yours."

"Then what the hell are you out here for?" yelled Daren, tired of dancing around his former leader's feelings. "You know what I do. You know what I'm about."

"Yeah, I know exactly what you're about," Marcus fired back.

"Yes, I should've said something. But let's be real, Marcus. You need this job as bad as I do."

"What job?" asked Marcus, finally realizing there was likely no pot of gold at the end of this rainbow. "You're only doing this to save your—"

"Your money is secure," said Daren. "I'm coughing up everything I've saved over the years. In fact, we can settle the account right now if you want."

But Marcus didn't respond.

"You're pissed," said Daren. "I get it. But damn-it, Marcus, this is it. We've got nowhere else to go."

Marcus couldn't believe the bind in which he now found himself. He sat down in the captain's chair and swiveled toward the main view port.

It was a tough choice given his own problems. *Take the money and run? Or help Daren clean up this mess?* And as he leaned back in the captain's chair, gazing through the view port into the vastness of space, he pondered long and hard over the next words that would come out of his mouth.

PART FOUR

THE DARK TEMPEST

"The Shadow People have awakened..." — Kalen

CHAPTER THIRTY-FOUR

Seven weeks ago, following the Maxis Lab job...

"Daren," shouted Tommy, hoping to snap their team leader out of his despondent sate. "We're wasting time."

Daren stared at Max as hot tears streamed from his eyes. He watched as Skye struggled to keep their dear friend from bleeding out on the deck of the gunship.

"Look at me," roared Tommy. "We have to complete the mission; it's our only play."

Daren looked at the black duffel bag containing the Benton Chamber, the device they stole from Maxis Labs at great cost. He closed his eyes and nodded. He hated that Tommy was right. They had to complete the exchange, or they'd be on the run for the rest of their lives.

Relieved to see that Daren hadn't taken leave of his senses, Tommy left the passenger bay, rejoining his copilot, Lana Doran, on the bridge. They failed the mission to kidnap the scientists, and Tommy knew they'd catch hell for that, *but at least they had the Chamber*, he thought.

"I heard screaming. What happened?" asked Lana.

"Just close your mouth and fly the ship."

"That girl needs help. Maybe we should—"

"I said shut up and fly the damned ship."

Lana complied, startled by a tone she was unaccustomed to hearing from Tommy. Her stomach ached as beads of sweat trickled from her brow. No longer concerned with the money, Lana wanted only to go home.

"Set a course for the Argos system, third planet," barked Tommy.

He could sense the distress radiating from his copilot but didn't care. They had a job to complete, and Tommy was desperate to cross that finish line, no matter the cost. Lana obeyed the order. She plotted the new course after bringing the ship to a complete stop. She pulled the acceleration lever. And with a sudden jolt of speed, they made the jump to hyperspace.

#

The gunship appeared in the target star system in a brilliant flash of light. Moments later, they breached the atmosphere of Argos like a flaming meteor. Once planet-side, Tommy navigated the ship toward a cluster of rundown buildings and landed the vessel. Before leaving the bridge, Tommy scanned the ship's surroundings. According to the readings, the planet was once a lush green world filled with life. But the surface was now a desolate wasteland, riddled with ruins, the result of a planet-wide nuclear war that occurred over two thousand years ago. No one survived.

#

As Tommy, Lana and Raven Squad filed from the rear of the starship toward a rundown office building near the landing zone, Daren lingered behind in the passenger bay with Skye and Max.

"She gonna make it?"

"Don't know." She looked at Max and sighed. "She's in an induced coma," she continued. "I had to inject her with quantum probes just to combat the swelling in her brain. But we need to get her out of here, fast."

"Trust me, I don't plan to be here a second longer than we have to," said Daren as he placed the Benton Chamber into a portable Mk17 storage unit. As he did so, the container morphed itself into a rectangular shape, just large enough to fit the device.

"What are you doing?" asked Skye.

"Just thinking ahead," Daren replied. He waved his hand over the top of the box, causing a holographic keypad to appear. He entered a code, and the locks engaged, causing the container to seal itself into a seamless block of dark metal. Daren stuffed the container back into the oversized duffel bag and exited the gunship, leaving Skye behind to care for Max. But as he started toward the rest of the team, Skye ran to meet him, an XR-17 rifle in hand.

"What about Max?" asked Daren.

"I can monitor her condition on my med-unit," said Skye, motioning toward the flexible view screen attached to her right forearm band. The device displayed Max's vitals and the status of the quantum probes. "Besides, we'll need every gun we have out there, in case this thing goes south."

Daren couldn't argue with her logic and made no objections.

#

Moments later, Daren and Skye entered the office. Everyone was there except for Tony, who had left the group to scout the surrounding area. Tommy stared at the duffel bag slung over Daren's shoulder, using his tactical eyewear to conduct a passive scan of its contents. The readings were strange. He needed to get closer, but Tony returned before he made his move.

"Saw a ship land nearby. I think they're here," said Tony.

Ten minutes later, Tommy received a silent alarm on the data pad built into his wristband.

"This is taking too long. I'll go meet them," said Tommy. He quickly exited the room.

The team exchange confused looks, but before they could say a word, the unspeakable happened. With a loud thudding sound, sonic grenades came flying through the broken windows, ricocheting off the walls and bouncing toward the center of the room. The grenades exploded with a loud disorienting shrill. The blast sent a blue concussive shock wave that sent everyone crashing into the walls. Jason and Lana hit their heads on the debris scattered throughout the room. They were knocked unconscious.

As the rest of the team staggered to their feet, Tony was hit in the torso with four high-powered shock rounds. Powerful jolts of electricity surged through his body, causing his cybernetic limbs to seize. He fell to the ground in a violent convulsion, then passed out.

Before Skye could react, they blasted her in the chest with a smoke canister, sending her once again colliding with the wall. Her body was racked with such pain, she couldn't move. Seconds later, the canister that hit Skye exploded, filling the room with thick smoke, obscuring everyone's vision. They gagged on the toxic vapors.

Daren tried to make his way to Skye, but one assailant struck him in the face with the butt of a rifle from out of nowhere. Skye pulled her pistol from its holster, but another attacker kicked the weapon from her hand. She looked up only to see the gloved fist that collided with her jaw, knocking her head against the stone floor. From her perspective, the entire room went black.

With their targets neutralized, the three armed assailants moved to the center of the room, still wearing the black balaclavas that protected them from the lingering gas in the air. They removed their masks. It was Tommy, flanked by two menacing Lyrians, Goran and his brother, Sarrus.

"This is stupid," yelled Goran, the shorter, stockier of the two Lyrians. "We should just waste them."

"I told you. We're better off keeping the crew alive," said Tommy while making his way to the black duffel bag lying next to Daren.

Tommy had explained to his two Lyrian accomplices that his employer, Teric Winters, was a paranoid man. If his second in command, Samson Kull, showed up to the meeting to find everyone slaughtered, they'd be hot on their trails before they could leave the star system. But if they kept them alive, Samson might jump to the conclusion that Daren and his crew staged the whole thing. Or, he'd waste time having Raven Squad beheaded for their monumental failure. Either way, Tommy figured the distraction would buy them enough time to disappear without a trace.

Though Goran and Sarrus could point out at least a half dozen flaws in Tommy's reasoning, they had no time to argue.

"They'll be here any minute," yelled Sarrus. "We're heading back to the ship, so hurry this up."

The two Lyrians left the building, leaving Tommy to claim their prize. But Tommy opened the duffel bag only to find that a solid metal container housed the Benton chamber. He scanned the box, but could find no opening, though his tactical shades confirmed the device's presence inside.

"What the hell?" roared Tommy. He grabbed the bag and donned his facemask.

Moments later, Daren overtook him after regaining consciousness. The two men wrestled on the ground, trading blows like gladiators locked in a perilous struggle of life and death. During the scuffle, Daren unmasked his opponent. When he saw Tommy's face, Daren lost it. He locked his hands around the traitor's throat. Unable to pry away Daren's hands, Tommy was moments away from passing out. He struggled to free his XP-90 handgun from the holster strapped to his leg. He worked the weapon free, jammed the barrel into his attacker's chest, and pulled the trigger.

The force of the blast sent Daren flying halfway across the room. Weak and gasping for air, Tommy looked over to see Daren lying motionless on the floor. While he knew Daren's advanced body armor would have absorbed the worst of the blast, he also knew the guy wouldn't be getting up anytime soon. Tommy staggered toward the Mk17 container, stuffed it back into the bag, and headed toward the exit.

"Tommy," cried a soft but disoriented voice.

He looked up to see Lana standing between him and the doorway, pressing her hand against the bleeding wound on the back of her head. Tommy hated that Lana had awakened. More than that, he regretted giving into her repeated pleas to go with them on the drop, as it would have been much easier to complete the deed without seeing her face.

"What's going on?" asked Lana, with tears streaming. "I... I just wanna go home."

"You will," said Tommy, struggling to find the words. He tried to tell her to go back to the ship, but she kept repeating she wanted to go home amid the flood of tears. But before Tommy could calm her down, he soon found himself covered in her blood, following the deafening sound of a pulse rifle blast.

Lana stumbled toward her lover in a lifeless heap, falling into his arms, her opened eyes frozen in time, the unnerving look of horror still etched upon her face. Tommy had seen many people die in his day, many by his own hand. But this one was different. This one *felt* different. He stared at her for a moment, then came to his senses, dropping Lana's body to the floor.

He looked past her to see Goran standing in the doorway with a rifle in hand, still smoking from the round that claimed Lana's life. Part of Tommy wanted to draw his own weapon and return the favor. But he could ill afford to show weakness in front of the Dragen Alliance.

"Get movin', Lance," yelled Goran. "Samson and his crew just jumped into the sector."

In his profession, there was no room for emotional attachments. All that mattered was the job. And Tommy did his best to live by that tenant. But in that moment, something changed for him. With the Mk17 storage container secure, Tommy tried his best to push the nagging thoughts of Lana from his mind. But before exiting the building, he took one last look at Lana's body, and at the devastation wrought upon Daren and his crew. He took a deep breath, turned his back to his former team, and followed Goran out the door.

CHAPTER THIRTY-FIVE

Protos IV: Present day
I just want to go home.

Ever since the day Goran pulled that trigger, Tommy couldn't escape the haunting sound of Lana's voice. Like his inability to alter the ebb and flow of the oceans on Protos IV, he was powerless to withstand the endless wave of thoughts crashing against every corner of his mind. Even the strident crackle of the enhanced Erecian cutting torch he used on the Mk17 could not compete with the unsettling words echoing inside his head.

Am I losing it? Tommy thought as he tried in vain to pierce the hardened outer casing of the container. Tommy often boasted of his complete *mastery over his emotions*. But now he questioned his own ability to keep his head in the game. A sharp pain radiated from the base of his skull to just behind his eyes. So, Tommy deactivated the torch and flung the expensive device across the room.

"You've been at this for weeks," said Goran from across the messy workshop. "I'm telling you; it won't work."

"Damn," yelled Tommy, exasperated by the unending mental distractions, and the fact he hadn't scratched the surface of the pitch-black cargo container. "The guy told me that torch could cut through anything," he said. "Wonder if it's too late to get my money back."

Sarrus approached Tommy's workbench. The scales on his face were not as pronounced as those on his elder brother, Goran. This rare birth defect caused many to question if Sarrus was Lyrian at all. In fact, some among their people thought Sarrus to be a Human-Lyrian hybrid, an abomination that should never have lived beyond birth. But he used his retractable gauntlet blades to disembowel those who gave voice to such sentiments. The Lyrians called the dual claws, *The Talons of Drakos,* named after the species of the dragon etched upon the crest that represented their family's house. Sarrus examined the surface of the Mk17.

"Not even a burn mark," said Sarrus. He rubbed his hand across the spot where Tommy focused the torch. "And still cool."

Tommy removed his tactical shades and rubbed his temples to ease the mounting pain. "What the hell is this thing made of?"

"I told you, it's an Mk17. It's impenetrable," said Goran. He rubbed the back of his heavily scaled head and sighed. "And did I mention, impossible to scythe?"

"Only a thousand times," muttered Tommy under his breath.

"And you left the only guy with the access codes, sound asleep on Argos Three," yelled Sarrus.

"How was I supposed to know he put it in a..."

"An inept human," said Sarrus. "Imagine that."

Tommy moved face to face with Sarrus, studying every line of his pale green face. He grew tired of their condescending, smug remarks. And for a moment, he was prepared to throw the whole plan out the window and give the two brothers what they had coming to them.

"Is there a problem, Lance?" said Sarrus, with a slight smirk on his face. He sported rows of razor-sharp teeth. With a flick of his wrists, he deployed the Talons of Drakos. The dual eighteen-inch blades, forged of ancient, serrated dragon bone, extended from his gauntlets with a loud shriek. The curved talons had a smooth finish, as if made of a highly burnished dark metal.

As Sarrus raised his arms into an attack position, Tommy could see intricate, luminous engravings written in ancient Lyrian script running the length of each talon. The craftsmanship was extraordinary, but Tommy was a man of simple tastes. So, he sprang back and pulled his XP-50 heavy assault handgun, aiming its sleek, matte-black barrel at the head of the Lyrian.

"Sorry, man," said Tommy, returning an arrogant smirk of his own. "I ain't into the medieval thing." He disengaged the weapon's safety, knowing his augmented reflexes would allow him to get a shot off well before Sarrus could mount an offensive.

Seeing that is brother was in trouble, Goran intervened, reaching for a specialty weapon of his own. With his right hand, he grabbed from his belt the bladeless hilt of an ancient two-handed long sword. Reaching out with his mind, Goran ignited the weapon, causing a blade of pure fire to extend from the hilt within a fraction of a second.

The Lyrians called the 500,000-year-old relic *The Breath of Drakos*.

When ignited, the weapon resembled the length and look of an ancient claymore, except it had a solid blade, breathed in pure dragon fire, burning hot with its signature greenish-black flame. With a powerful downward stroke, he severed the hardened barrel of the handgun with unexpected speed and finesse. The scorching heat of the sword forced Tommy to recoil his hand in pain. The temperature emanating from the blade seemed almost unnatural. Goran then delivered a powerful kick to Tommy's chest, sending him crashing to the floor. From the ground, Tommy reached for his secondary weapon, but Goran was upon him before he could touch the grip of his XP-90 tactical pistol. Goran moved forward, pointing the business end of his sword into Tommy's face.

Though the blade was over twelve inches away, Tommy felt as if his skin were on fire. He groaned in pain, having never felt such withering heat, not even from the most powerful Thermo-Blade in existence. He squirmed to move away, but Goran closed the gap.

"You are here only because I found you to be, *useful*," snarled Goran. He moved the blade two inches closer, causing Tommy to yell in extreme pain. "But it seems your usefulness has run its course," said Goran, motioning toward the device that still housed their coveted prize.

Goran deactivated the blade, causing it to dissipate in a greenish-black haze. He returned the hilt to his belt and moved toward the Mk17, casting a stern glare upon the indestructible apparatus. "We need to update Ar'Gallious regarding your... lack of results."

"Come on, guys," pleaded Tommy, staggering to his feet. "Ain't nothin' impenetrable."

"Is that so?" asked Goran. With blinding speed, he reached for his sword, once again igniting its ominous blade of fire. And with all his strength and fury, Goran struck the Mk17 with the edge of his burning blade, sending it flying across the room, where it imbedded itself into a solid metal wall.

Goran wasn't worried about damage to the Benton Chamber, knowing the Mk17's built-in kinetic stabilizers protected its contents from any external force.

He grabbed Tommy by the throat and marched him across the room, then he released the mercenary and yanked the container from the wall, causing it to slam to the floor with a resounding thud. Goran deactivated the weapon, returning the hilt to his belt.

"Look at it, Lance. Not a scratch!" yelled Goran, pointing toward the pristine storage unit. The container was still smoking from the powerful strike of his sword. "Any other container *would've been split in two* by this weapon," he roared. "You *can't* open it without the access codes."

Tommy believed him, having seen firsthand the ease at which Goran's blade ripped through his gun, which was impervious to heat and flame. He then studied Goran's eyes, seeing a fire within them that burned hotter than the Breath of Drakos itself. *These guys mean business*, Tommy thought. He took a deep breath and stepped back.

"I can get into it," said Tommy. "You gotta trust me. But I need more time."

Goran ignored the babbling mercenary. He walked across the room to join his brother near the computer terminals. Before Tommy could speak again, the two Lyrians exited through the sliding metal door to contact their illustrious leader.

Tommy returned the Mk17 to the workbench, then paced the floor of the workshop for a moment, realizing that if he didn't crack the case on this thing soon, they would cut him out of the deal altogether. And by *cut out*, he meant shot and left to bleed out on that backwater, annoyingly colorful planet. Tommy walked across the room to the computer terminal upon which Goran had been working. He examined the last search that Goran had made. To his surprise, he saw information about the Mk17 container manufacturer, Galaxy Tektronics. Tommy conducted further research and smiled, realizing that he may have just found a kink in the cargo container's so-called *impenetrable* armor.

CHAPTER THIRTY-SIX

The silence was absolutely deafening as Daren watched Marcus stare into the void of space through *The Indicator*'s forward view screen. While Marcus weighed his options, Daren kept his mouth shut. Over thirty minutes of agonizing stillness had passed. Though desperate to know his decision, Daren reasoned that Marcus had earned a few moments of quietude. After all, he had just complicated his brother's life in ways that were almost unfathomable.

As Daren sat in the copilot's chair, he fidgeted with the nonessential flight controls on the largely holographic instrument panel. However, his mind continued to regurgitate all he had done, despite his best efforts to suppress those thoughts. But the staggering list of atrocities committed by his own hand weighed so heavily upon his heart, Daren could experience no mental reprieve. And in an instant, the thoughts came flooding through his mind like a breaking dam.

His reckless ambitions had caused their dear friend and former crewmate, Max, to end up in some rundown medical facility, fighting for her life. Because of his lies, Daren caused Marcus to endure an attempted mutiny aboard his own ship while involving him with stolen government technology that carried with it the potential to plunge entire civilizations into war on a galactic scale. As the endless stream of thoughts continued to fill his mind, Daren rose from his seat. He paced the metallic deck of the spacious bridge. But Marcus continued to sit in his chair, as if he were the only person in the room.

Daren thought he knew everything about Marcus, having grown up alongside him since their early teenage years. Through the harsh mercenary lessons imparted to them by Daren's father, Jax, there was forged between Marcus and Daren an almost perfect and unbreakable bond. And from that bond, they became more than friends brought together by a mere circumstance. They became brothers. Yet despite their history, Daren was unable to guess what Marcus was thinking as his friend and former leader continued to stare into space out the view screen.

Daren continued to pace the steel deck of the dimly lit bridge, thinking upon the countless times he selfishly *complicated* Marcus' life. But through it all he was certain the connection they shared would always endure, like the endless ocean of stars that formed the majestic galaxy through which they traveled. But as he stood watching Marcus wrestle with a decision he should have never had to make, Daren found himself standing upon unfamiliar ground. And now he would have to face the thing he refused to acknowledge, the truth that the unbreakable was now broken, and that things would never be the same. And just when the weight of it all had become too much for Daren to shoulder, the deep voice of Marcus cut through the bitter silence.

"No," said Marcus, "I can't help you."

Daren felt like his brother ripped his beating heart from his chest. He staggered back to the copilot's chair and took a seat, unable to utter a word.

"Ever since I've known you, it seems like all I've done is drag you behind me, like a dead weight," said Marcus. He turned in his chair to face Daren.

Marcus recounted the many times he had stepped in to rescue Daren from the deadly jaws of his own asinine decisions. The track record was long and difficult to hear. But Marcus continued, figuring that it was time to speak that which had remained unspoken for far too long.

During his heated tirade, Marcus spoke of the time when Jax had given Daren his first big break, entrusting him with a valuable arms shipment he was to sell to a shadow market dealer calling himself Barrek. However, the job went horribly wrong when Barrek's girlfriend seduced and then swindled Daren out of his cargo. The same woman Daren had met and fell for just two days earlier had been playing him from the beginning.

"After that mess with Barrek," said Marcus, continuing his discourse, "things were so bad between you and your dad..."

"Yeah, I remember," shouted Daren. He refused to let his mind revisit that dark place. "I don't need a reminder."

But Marcus continued.

"It was *me* that kept you from putting a round through your head that day," said Marcus. "*I* took the heat from your father; all to shield you from *your own* stupidity!"

Daren sat in silence. There was nothing he could say. Marcus could have continued, but there was no need. He made his point. He had been there for Daren on countless occasions, never asking for anything in return.

"All you do is take and take," said Marcus, reaching the apex of his frustrations. "But it ends here." He looked into Daren's eyes, as if searching the depths of his soul. "I went to prison for ten years, Daren," said Marcus. "All to clean up what you screwed up."

Marcus paused. He had run out of words. They regarded each other uncomfortably for a moment, then Daren spoke.

"I know," said Daren. "But I promise, things will be different this time—"

"No more promises," said Marcus. He took a deep breath, then continued. "But a deal is a deal. So, I'll honor our contract and take you to Protos IV. But upon arrival, you will pay me in full. No negotiations. No excuses."

Daren agreed, relieved that Marcus didn't call the whole thing off and dump them on the nearest planet.

"But I'm not helping you get your cargo back," said Marcus. "It's time you stepped up and put in your own work for a change." Marcus returned to his seat, swiveling in his chair to face the command console. "When we arrive, you have twenty-four hours to handle your business. Or, I promise, I'll leave you on that planet."

Twenty-four hours was a tight timeline. They had tons of reconnaissance and prep work ahead of them before they could move on Tommy's safe house. But Daren made no protest.

"Thank you, Marcus," said Daren. "You don't know how much this means."

But there was only icy silence. He rose from his chair and headed toward the exit. But before crossing the threshold of the bulkhead door, Daren turned to look once more at his brother. But Marcus never turned his gaze from the console. Unable to find words to express his remorse for all he had done, Daren let out a slow, audible sigh. He then turned and exited the bridge, in utter disgrace.

#

Within the hour, *The Indicator* exited hyperspace on a fast approach to Protos IV. From space, the view of Protos was nothing short of breathtaking. The planet rotated on its axis with the look of a resplendent opal. Its vibrant colors shimmered brilliantly in the light of its giant yellow star, around which seven other planets circled. Because of its unique soil composition, the planetary governor dedicated the entire world to grow rare and exotic crops. As a result, farmers from across the galaxy paid exorbitant sums for plots of farmland, hoping to build their agricultural empires.

On the surface, there were no sprawling cities or massive settlements. There were only small villages, large enough to house the farming staff that controlled the mostly automated labor force across each continent. As Marcus scanned the surface, he could see why Tommy chose this location for his hideout. He figured that such a sparsely populated farming planet, in the middle of nowhere, would be the last place ISG military forces would look for hardened criminals.

After placing *The Indicator* into a polar orbit of Protos IV, Marcus put the ship into stealth mode and locked onto the surface coordinates provided to him by Daren. After starting the ship's auto landing sequence, he joined the crew in the cargo bay, where he approached Daren.

"Get your crew ready, we'll be landing soon," said Marcus. He lowered his voice so that only Daren could hear. "I'm not playin' with you, D, you have twenty-four hours. And the clock starts as soon as we hit the surface."

Daren acknowledged and ordered his crew to ready their weapons and gear. As they did so, Skye approached Marcus just before he headed back to the bridge. Skye still had much she wanted to say, and given the possibility they may not survive the encounter ahead of them, she figured it was now or never.

"I don't suppose you'd change your mind about coming with us, huh?" asked Skye with a slight smile.

Marcus paused. Though he tried to deny it, he had deep feelings for her that neither time nor space could ever diminish. Part of him hated leaving her life in the hands of Daren. But things had changed for him.

"I can't, Skyela," said Marcus. "I have a family to think about now."

"*We* were once your family," said Skye. "But I get it. Your *real* family needs you back home. I mean, who can be mad at that, right?"

Skye's words cut deeply. After practically severing ties with his biological family, the members of Raven squad were the only family he had acknowledged for years. With such a long history between them, his decision not to join them on the mission caused his heart to sink as if bound by a massive weight. As he watched her, memories of the times they spent together flooded his mind. Though he tried to suppress it, Marcus couldn't deny the fact that he still loved her. And though the circumstances were far from ideal, reuniting with Skye was something that he longed for, especially during the long years he spent locked up at Runner's End. As Marcus looked intently into her eyes, he gently grabbed her by the hand.

"You know," Marcus said, trying desperately to find the words. "You don't have to do this."

"C'mon, Marcus," said Skye. She wanted nothing more than to drop everything and run off with him. But she had a job to do. She sighed, then spoke. "You know how this works."

Marcus knew the game all too well. Teric Winters wasn't the type to let things slide. The mission had to move forward.

"Well, maybe when this is over..." said Marcus, stopping himself before finishing the words.

As the moments passed, Skye waited expectantly for him to complete the words that she longed to hear. But the words never came. In his mind, there was no way Marcus could live in both worlds. There was no way he could care for his family back on Stratus One while at the same time holding onto his past. And as much as he hated it, Skye was very much a part of that past that he needed to let go.

"Never mind," said Marcus. And without saying another word, he turned and quickly left the room.

Skye stood silently watching the bulkhead door as it slid shut behind the man that she had once vowed to be with until the end of her days. She shook her head in disappointment. For she too loved him deeply. But as she stood staring at the closed door she began to think, *Maybe it is time to move on.* She wiped away the tear welling in the corner of her eye. There was so much she still needed to tell him. But in that moment, it was apparent that she wouldn't have the chance to say the things that she kept bottled inside for more than a decade. So, she slowly turned and rejoined the crew. She had to keep her head in the game, knowing that they stood moments away from embarking on a mission that could very well be the final chapter for their old crew, Raven Squad.

CHAPTER THIRTY-SEVEN

"Captain Shepard, we'll be docking with Outpost Theta in twenty minutes," crackled the shuttle pilot's voice over the ship's intercom.

Donald opened his eyes upon hearing the pilot's voice and stood up to stretch. The passenger bay featured several rows of uncomfortable seating and a few overhead storage compartments. He was fortunate to be traveling alone, as there was little to no space between seats. He approached the forward bulkhead that separated the bridge from the passenger bay.

"Copy that," said Donald after activating the comm unit affixed to the bulkhead.

He retrieved his rucksack from one of the overhead compartments and prepped his gear. Having spent many sleepless nights in search of the stolen Benton Chamber, which had become the bane of his existence, Donald used the lengthy shuttle ride to the Outer Core as an opportunity to catch up on some much-needed rack time. He knew that all the higher-ups were watching, including President Jonathan Vance himself. So, the last thing he wanted was to arrive to Outpost Theta only to find their intelligence to be flawed. But Donald pushed those thoughts from his mind. The elite soldiers of Saber Team Seven were the highest tier operators of the entire saber brigade. If his team said they had found the safe house, then Donald would stake his reputation, his career, and even his life on their word.

Donald joined the pilot inside the compact bridge and took a seat in the copilot's chair. With a few presses and swipes of the navigational controls, he pulled up a better view of Outpost Theta. The cross-shaped weapons platform resembled a massive dagger floating in space. Armed with countless pulse cannons and rail-guns, the outpost also hosted an array of offensive and defensive missile systems. Besides its impressive armament, Outpost Theta had sizable hangar bays attached to its underbelly, capable of launching thousands of manned and unmanned fighter-squadrons at a moment's notice. One look at the menacing superstructure and it was clear the outpost had one purpose, to wage war against the enemies of the ISL.

Being an avid history buff, Donald reflected on the old stories regarding the gargantuan outposts. It was hundreds of years ago, during the time of the Great Purge, that the Interstellar League commissioned eight deep-space battle stations. These massive, hyper-jump-capable weapons platforms served as mobile staging points from which the Interstellar Guard launched surgical strikes against Orion's Shield. Many historians believed it was those platforms that turned the tide of that long and bloody civil war. After the Great Purge, the ISL decommissioned all but the eighth and most advanced station, *Military Outpost Theta*.

Donald had been to the station many times in the past, but only as an instructor, helping to train shadow tech veterans in the fine art of advanced intelligence gathering. *But one thing is for certain,* Donald thought as he reviewed the preliminary reports sent to him from his team, *lesson time is over, and things are about to get real.*

#

The hangar doors slammed shut as Donald's shuttle entered the Outpost. Within minutes, the ship touched down on landing pad fourteen, after which the pilot cut the main engines. As the aft ramp of the craft extended to the floor, Donald exited the shuttle where his second in command, Lieutenant Alex Chavez, awaited him on the flight deck.

Alex stood six-foot-three, weighing in at two hundred thirty pounds. All muscle. He had strong facial features that seemed almost chiseled in stone. He was a soldier's soldier, devoted to the job, with few family ties. But there was more to Alex than his impressive physique and unmatched fervor. He possessed all the qualities the ISG looked for in young officers, but seldom found in one package. And with the aid of his eidetic memory, a keen mind for strategy, and natural leadership skills, Alex graduated at the top of his class at the Saber Brigade Officer Academy. While ever grateful to have Alex as his right hand, Donald knew it was only a matter of time before his protégé received his own command, even though he swore he would never leave the team. In fact, Donald had to twist his arm to become an officer in the first place, which Alex only accepted under the condition he could return to Saber Team Seven to serve as assistant team leader.

"Talk to me," said Donald as he and Alex walked past the sea of fighter craft uniformly parked throughout the multitier hangar bay.

"Our scouts on the surface reported in," said Alex. "We have it, sir."

"Positive?" Donald asked, trying to mask the excitement in his voice.

"Yes, sir," Alex replied with a slight smile breaking the corner of his mouth. "We're picking up trace elements of the Ladium crystals inside the Benton chamber."

"Anything on audio?" said Donald.

"Audio surveillance confirms the prototype is on site. But it's weird. Sounds like some merc they double-crossed stuck the chamber in an Mk17 storage unit before it was taken," Alex said, chuckling. "They can't even open it."

"Whoever said there was honor among thieves clearly hadn't met these guys," said Donald as he reviewed the holographic transcripts on the comm unit embedded in his left wristband.

"Recon of the area shows they're dug in pretty good," continued Alex, "but if we're gonna move, we'd best do it now."

"Agreed," said Donald, deactivating his wrist pad. "Assemble the team," he continued. "I want to be on the surface of Protos IV in less than two hours."

Alex's eyes widened with excitement, knowing that soon they'd be back where he and the rest of the team were most comfortable, on the field of battle.

As he watched Alex run off to prepare the team, Donald breathed a sigh of relief, knowing that his idea to scan for Ladium had paid off. They caught a huge break when they learned that the research team had been conducting small-scale tests of the prototype before its theft from Maxis Labs, and that it was likely that small fragments of the crystals may still be inside the chamber.

As impressive as the Benton chamber was, the device itself was little more than a means of harnessing the immense energy output of those incredibly rare Ladium crystals, intended to be the replacement energy source for the failing Krillium-powered terraformers. But the best part about Ladium was that it possessed a unique energy signature, unlike anything found in the known galaxy, making it easy to spot with properly tuned scanners.

Though General Kirkland had urged him to run the entire operation from Outpost Theta, Donald had no intentions of staying behind. There was too much at stake, and he wouldn't be satisfied until he saw the Benton chamber for himself, regardless of what the intelligence confirmed. As Donald continued from the hangar toward the lift that would take him to the command center, he felt like a child on the eve of his birthday, brimming with exhilaration. Yet despite his excitement, Donald contained himself. But still, he was pleased to know that all would soon be made right, once they finally made planetfall.

CHAPTER THIRTY-EIGHT

Protos IV...

Night had overtaken the land, carrying with it a warm and gentle breeze, including the subtle alien sounds of the nocturnal beasts as they emerged from their hiding places. The planet's flora was like something out of a fairy tale, with vibrant colors permeating its breathtaking landscape.

As he traversed the terrain, Jason left faint boot prints in the fertile, teal-colored soil for which Protos IV was so famous. Yet as beautiful as his surroundings were, Jason never considered himself to be the outdoorsy type like his sister, Skye. His idea of a good time was spending all day behind the flight controls of a starship, or neck deep in its innards, installing the latest in his impressive line of custom upgrades. So, when Daren tasked him with scouting the defenses of Tommy's hideout, he wasn't excited about taking the job, knowing he'd have to traverse a densely wooded area just to make it to the target location. But they were a small crew, and everyone had a part to play, so he accepted the task without protest.

Beyond the forest tree line where he stopped to observe Tommy's hideout, Jason saw an amber-colored beach stretching as far as the eye could see. Scattered throughout the yellowish-brown sands were a host of multicolored crystalline formations protruding from the ground like clusters of serrated teeth. At the center of the picturesque beach two miles from the tree line stood Tommy's massive safe house, which looked more like a stone fortress than anything else.

For more than two hours, Jason scanned the stronghold's rear defenses. He used advanced electronic binoculars that fed the data to the computer embedded in his left wristband. And for a better look at the front of the structure, as well as on the roof, Jason deployed a small stealth drone he and Tony designed a few years back, small enough to evade all active and passive sensor sweeps. As the drone flew silently above, it tagged every guard and sentry cannon on and around the structure.

Whoever Tommy's working for, must be paying well, Jason thought, surprised by the level of security and the sheer number of hired guns he'd been tracking in and out of the facility. After cataloging the structure's defenses, Jason recalled his drone, packed up his surveillance gear, and began the three-mile trek back to the clearing where Marcus had landed *The Indicator*. As he made his way through the dense forest, Jason tried to suppress the gnawing feeling eating away at the back of his mind... the feeling that they may very well be in over their heads.

#

About forty minutes later, Jason emerged from the dense tree line into the clearing where the team had set up a small base camp. As he made his way toward the crew, Daren moved forward to meet him.

"Looks like Tommy's doin' well for himself," said Jason while transferring the data from his wristband to Daren's data pad. "The guy's got a small army." As Daren silently studied the surveillance data, Jason dropped the rest of his gear to the ground. "That place is on lockdown for real," continued Jason. "Hell, I know some of those guys. And they're definitely not folks you want to dance with."

Moments later, Tony and Skye stepped forward.

"Any chance we could convince them to join our little crusade?" Tony asked Jason as he looked over Daren's shoulder at the data pad.

"Those guys would sell their souls if they could get enough for it," said Jason. "But they're way out of our price range."

Jack, still sore from his run-in with Marcus during his attempted mutiny less than fifteen hours earlier, joined the small group after overhearing the conversation. "Just gets better and better," said Jack while glaring at Daren.

"We can do this," said Daren without lifting his eyes from the data pad. "We have to," he said, finally fixing his eyes on Jack. "You wanna walk? Be my guest." Daren picked up an empty rucksack from their supply stash and shoved it in Jack's chest. "If not, then shut up and gear up."

Jack snatched the bag from Daren and stormed off to ready his gear. As Daren and Tony discussed the infiltration plan, Jason pulled Skye to the side.

"Check this out, Skye," Jason said, struggling with how to begin. "I need you to sit this one out. You know, just hang back at the ship with Marcus."

"*Sit this one out?*" said Skye, slightly amused. "Have you lost your mind?"

"Look, you have no idea what we're up against out there."

"Yeah, Jay, I *do* know. I've seen the surveillance."

"Come on, Skye. I admire your courage and all, but now ain't the time for the 'strong woman' routine."

"Excuse me?" said Skye, offended by her brother's remark.

"Those guys out there," said Jason, pointing toward Tommy's safe house, "they're animals. Especially the Lyrians. And they don't take prisoners."

Skye dismissed her brother's words and walked off, but Jason grabbed her by the arm, spinning her around to face him.

"Why do you always have to be so difficult, girl?" Jason said, yelling loud enough to make a scene.

"*Girl?*" said Skye, raising her voice in kind. "Let me tell you somethin'... As big as you are, I'm still *your* older sister," she said while wrenching her arm free of his grasp. "And in case you forgot, I'm *more* than capable of holdin' my own."

"Yeah, but it ain't about that," yelled Jason. "There's no sense in all of us gettin' killed over this dammed box," he continued. He paused for a moment and took a deep breath. "I'm just tryin' to protect you."

"I don't always need you in front of me, Jay," yelled Skye in frustration. But she knew he meant well. So, she paused for a moment to calm herself, then lowered her voice. "Sometimes... I just need you beside me."

The two remained silent for a moment as Jason took in his sister's words. It was bad enough that Marcus remained on the ship, having decided not to help after his fallout with Daren. And Jason knew they couldn't afford to lose another gun. So, he conceded, deciding not to press the issue further.

"Now come on, boy," said Skye after punching her brother on the shoulder. "We've got a party to crash."

As Skye ran off to grab her gear, Daren called to Jason.

"We good over there, Jay?" said Daren after overhearing much of their conversation. "It's almost go time."

"Yeah, I'm good," said Jason, "I'll be there in a minute."

As he observed Skye readying her weapons, Jason knew there would be no talking her out of this one. So, he instead grabbed his equipment pack and moved by his sister's side, resolving within himself that whatever they were to face, they would face it together.

#

On the bridge of *The Indicator*, the thoughts of Skye still weighed heavily on Marcus' mind. He wanted nothing more than to march out there and pull her away from that mission, which he felt was more of an exercise in foolishness than anything else. But he knew Skye could take care of herself, so he shifted his thoughts to other matters.

Marcus sat in the captain's chair, watching the holographic timer that he had synchronized with Daren's data pad just twelve hours earlier. He was serious about the twenty-four-hour time limit he imposed, though in the back of his mind, he had no desire to maroon his old team on the planet's surface. As the minutes faded away, Marcus found himself becoming increasingly unsettled. Perhaps it was because, in moments like these, he was used to being in control. Calling every shot. Overseeing every detail. But things were different now. He was just the cargo pilot on this job, and nothing else.

While he never considered Daren to be inept when it came to matters of combat, Marcus did question Daren's ability to keep his emotions in check, especially with all that had transpired in recent weeks. There were just too many variables in play; variables that he was sure Daren hadn't even considered. But Marcus had to stop himself from entertaining such thoughts. After all, the welfare of the team was no longer his responsibility, and he was good with that.

As the uneasiness grew, Marcus tried to keep himself busy, performing mostly unnecessary maintenance throughout the ship. But no matter what tedious chore he assigned himself, he was unable to keep his mind from wandering, and the anxiety from building. After running out of frivolous tasks to undertake, Marcus decided it was time to check in with his family. Using a high-tech deep space transmitter designed by Jason years ago, Marcus sent a secure audio-only transmission to his home on the Stratus One Star Port. Without the covert comm unit, Marcus would have been unable to send the signal without it being intercepted and traced back to the ship. Within minutes, he heard the reassuring voice of his sister on the other end.

"Well hey there, Marcus," said Samantha. "What's with the visual feed? I can't see you."

"Solar flares," said Marcus, lying through his teeth. "It's been wreaking havoc on comms since we got here."

But since his release, Samantha had become adept at not just reading her brother's face, but his voice as well.

"Is everything alright?" said Samantha, sensing the uneasiness in his voice. "How's the job going? You should be on your way back by now, right?"

"Daren said their contact was delayed," said Marcus, trying to maintain the lie. "But we should be out of here in about twelve hours."

"Twelve hours?" said Samantha, disappointed. "So, I guess we'll be seeing you in two days then."

Samantha had an uneasy feeling about that job from the start, and she wouldn't rest easy until those mercenaries were off that ship, and her brother was safely en route back home.

"Careful, Sam," said Marcus. "It almost sounds like you miss me."

"Yeah, right," Samantha joked. But before she could say another word, Elizabeth and their cousin Kayla entered her office, interrupting the conversation.

"Hey, Uncle Marcus," said Elizabeth. "You need to get here so we can make that trip to Earth. You blew me off once, but it ain't happenin' again," she said, followed by giggles from her and Kayla.

"I do have a job, you know," said Marcus, with a slight smile breaking the corner of his mouth. "But don't worry. I got you," he said. But this time he meant it. He had been selfishly withdrawn for so long that he had almost forgotten about the things that truly mattered.

"And I'm coming too," Kayla interjected.

"Is that right?" asked Marcus, amused that his cousin had suddenly invited herself on their trip.

"Yeah," said Kayla. "Unlike you, I'm actually *from* Earth. And without me, the two of you wouldn't have the slightest idea what to do after the university tours. Consider it a public service."

"Public service?" said Marcus, chuckling. "You sure it has nothing to do with the fact that your ISG boyfriend was reassigned to Earth?"

Kayla was glad Marcus couldn't see the embarrassment on her face, but every word he spoke was true. She had fallen for one of the soldiers garrisoned at Stratus One a few months back. And just when things were getting serious, her boyfriend dropped the bombshell announcement that he was being redeployed to Earth, where he was to join the Citadel Guard in Haven City, a most prestigious post for any soldier. So, Kayla had been looking for an excuse to visit Earth ever since.

"Yeah, that too," said Kayla, deciding to come clean. "But trust me, we'll have fun. So, can I come?"

"Of course," said Marcus without hesitation.

"Yes!" said Elizabeth, excited that her cousin was officially joining them on their upcoming trip to the Home World. "Well, we gotta go. Kayla is taking me shopping. I can't wear just anything to Earth, you know."

"I'm sure you can't," said Marcus. "I'll see you soon."

After Kayla and Elizabeth left the room, Samantha returned to the comm unit.

"Don't let them down," said Samantha. "I think this trip will be good for all of you."

"Yeah, I think you're right," said Marcus. "And don't worry, I won't let any of you down again. I promise."

Samantha had no idea what brought about the sudden change of heart for Marcus. But she was glad to see that he was finally turning a corner.

"Good. Now you hurry back," said Samantha. "It's not safe to be anywhere near Lyrian space right now with everything going on in the news."

"Why? What are they saying?" asked Marcus. He had purposely stayed away from the news for the last few days, though he already had an idea of what she was talking about.

"Well, to let Bobby Wiseman and the rest of those melodramatic reporters tell it, talks between President Vance and the Lyrians are falling apart," said Samantha. "They seem to think a Lyrian invasion is coming. All because of that device that went missing from Maxis Labs a few weeks ago."

"They've been saying that for weeks," said Marcus, trying to downplay the reports.

"I don't know, Marcus. It feels different this time."

"What do you mean?"

"The ISG Garrison here is now on high alert," said Samantha, her tone sounding almost ominous. "And the I.S.F. Onyx has just docked with the station this morning."

"What's strange about that?" asked Marcus. "Military ships come and go all the time."

"The Onyx is the flagship of the Interstellar Fleet," explained Samantha, forgetting that her brother wasn't as well versed in the inner workings of the ISL military. "But know this, that ship only leaves the Sol system on two occasions, when we are on the verge of war, or neck deep in it."

Marcus felt as if his heart was suddenly lodged in his throat. He knew the entire situation was bad, but after hearing that the most advanced warship in the ISL arsenal was now docked at his home star port, things suddenly became very real for him.

"Don't worry," said Marcus, trying to keep his sister from going into panic mode. "With something that serious, I'm sure the government is already on top of it."

"Yeah, I'm sure they are," said Samantha, deciding to take her brother's advice. "I'll tell you what. I'll stop worryin' the moment I see you walk through that door. Deal?"

"Guess I'd better hurry back then," said Marcus, touching the blank view screen in front of him.

"Okay then," said Samantha after a slow exhale. "I suppose it's time for me to get back to work. You take care of yourself out there," she continued, touching the blank screen in front of her as well.

"I always do," he replied.

After exchanging parting words, Marcus deactivated the audio transmission. He sat back in his chair, feeling as if his own stress levels were about to reach critical mass. He stood up and began pacing the deck of *The Indicator*. In truth, he had no reason to believe the ISL government even knew where the Benton Chamber was. And even if he leaked the intel to them, they probably wouldn't arrive before Daren and his team made their move on the safe house. The stakes were high, and on top of that, he had the sinking feeling that his former team was about to walk into a hurricane of epic proportions.

Marcus felt like he needed to do something, though he reasoned that at the end of the day, the presence of another shooter would make little difference. But on the other hand, he knew that if Daren and his crew were to fail, the potential impact on his family would be catastrophic. And that was something he would never allow.

He moved to a large compartment built into a wall panel located at the rear of the bridge. The partition was designed to be completely invisible to external sensor sweeps, perfect for the storage of illicit cargo and other items they wished to remain off radar. He opened the storage compartment to see the black tactical gear that he once wore back in his mercenary days. The very sight of it brought back memories both exhilarating and painful. Behind the dusty garments, he also kept several high-powered rifles and handguns in case he was ever boarded by pirates during his lengthy cargo runs to the Outer Core.

As he stood staring at his weapons and gear, his head was telling him not to get involved. But all he could think about was the well-being of Skye and the rest of the crew. He felt torn between the needs of his family back on Stratus One, and the needs of his family there on Protos IV. But after that last conversation with his sister, his gut was telling him that simply sitting on the sidelines may no longer be an option.

CHAPTER THIRTY-NINE

In total silence, Daren and his crew approached the tree line where Jason had conducted his surveillance, two miles from the rear of Tommy's safe house. With Jack on point, he raised his hand to signal for the team to come to a stop so he could move forward to scout the area. After observing the changes in the guard's patterns, Jack moved back toward the team.

"Looks like they set up a break area in the rear. They don't seem on high alert, but there's a bunch of them back there."

"Good," said Daren. "Maybe we can catch 'em by surprise."

"I'm not feelin' this," said Jack after seeing the building's defenses for himself. "How the hell are the five of us supposed to—"

"Six actually," said Marcus, to everyone's surprise, as he emerged from the densely wooded area just behind the team.

Dressed in full tactical gear, Marcus stepped forward armed with his trusty XR-17 assault rifle, and two XP-90 combat handguns. Marcus also donned a backup pair of the tactical eyewear with the purple tint, which he hadn't used since the Titan job eleven years earlier.

Daren let out a sigh of relief and approached Marcus. "Thanks, man."

But Marcus walked past him toward the rest of the team, without so much as looking Daren's way.

As he stood ready to address the crew, Marcus briefly locked eyes with Skye and smiled. There was no way he was leaving her life, or the life of his old crew, in the hands of Daren. As well versed in the art of warfare as Daren was, Marcus knew that his former second in command was way out of his league on this one. In her heart, Skye knew that Marcus wouldn't abandon her or the rest of the crew. She returned his smile with one of her own, excited that she might get to have that long overdue conversation with her true love.

"Everybody listen up," said Marcus. "If this is goin' down, then we do it my way." He turned to look at Daren, then Jack. "If anyone has a problem, then I take my ball and go home. Got it?"

Because no one wished to be stranded on Protos IV without a ride, there were no objections.

"I got it all mapped out," said Daren as he moved toward Marcus. But before he could continue, Marcus cut him off.

"Given your track record, I'm gonna have to call an audible on this one," said Marcus. He ordered the team to fall back a half mile into the woods from the direction they came so he could address the team. When they reached the new staging point, Marcus ordered Jason to transfer the surveillance data to Marcus' wrist pad. After studying the data for a moment, Marcus addressed the team.

"You know, part of me wanted to collect my money and leave you guys to figure this out on your own, when something occurred to me."

"We don't have time for this," said Jack.

But Marcus continued, ignoring Jack's words.

"All the family I have left is on the Stratus One Star Port," said Marcus.

"What does that have to do with anything?" asked Jack, irritated.

"Because Stratus One will be the first target hit if you guys screw this up, and the ISL ends up at war with the Lyrian Empire," said Marcus in a cold tone. "So, you'll forgive me if I choose not to leave the fate of my family in *your* hands."

But Daren knew where this was headed, and he didn't like it. In his mind, turning the Benton chamber over to Teric was the only way to get him off their backs for good.

"Come on, Marcus," said Daren. "Let's think about this for a moment."

"I already have," said Marcus, "I thought about my niece, my sister, and my cousin. So, when we get the Benton Chamber, we turn it over to the ISL. Period."

"You don't understand," said Daren, trying not to raise his voice. "Teric is obsessed with that thing, if we don't—"

"Teric can go to hell," said Marcus, shutting the conversation down. "And you can tell him I said so."

Following an uncomfortable silence, there were no objections, but Daren was furious that Marcus was changing the entire plan. Marcus knew how Daren felt, but didn't care. In that moment, his only focus was to do whatever he could to protect his family. Marcus signaled for the team to gather around. And as he laid out the new infiltration plan, Daren stood frustrated on the sideline with an icy glare fixed on their new leader.

#

Inside the dusty communications room of Tommy's safe house, Goran and Sarrus had been conferring with Ar'Gallious, leader of the Dragen Alliance, for over an hour. Believing Tommy's antiquated comm system wasn't up to the task, they opted to use an advanced deep-space transmitter designed by Sarrus. Not only was the transmitter capable of applying an advanced form of encryption to the signal, but it could mask the transmission itself, making it indistinguishable from naturally occurring radio emissions found throughout outer space. Sarrus had determined that doing a full holo-trans session would have increased the data size beyond their ability to conceal, so he compensated by sending an audio-only transmission.

"The time is near," said Ar'Gallious to his two most trusted and willing servants. The compression of the signal gave Ar'Gallious' deep and boisterous voice a filtered and more subdued tone, as if he were speaking through an armor-plated facemask. "Soon, the Lyrian Empire will be at war with the Interstellar League."

"And amid the chaos, we strike at the heart of the Empire to reclaim that which is ours," said Sarrus, echoing the words spoken to them by Ar'Gallious for as long as he could remember.

Being twenty-five years younger than Goran, Sarrus had only known life as an exile. He was but a child when Grand Warlord Ar'Gallious and his followers were expelled from the Lyrian Empire, following an attempted military coup to overthrow Emperor Ja'Gren, the father of the current Lyrian Emperor, Lord Zek'ren.

"And what of the mark seventeen container holding the Benton chamber?" asked Goran. "The mercenary called Daren West is the only one with the access codes."

"Ah, the son of Jax West," said Ar'Gallious. He had much respect for Daren's father, the legendary pirate and gunrunner who smuggled arms for the Dragen Alliance for years, unbeknownst to Teric Winters, and even his own son. "You will bring Daren and the container to me," said Ar'Gallious in his thunderous voice. "I will not be denied this glory. Not when we are this close."

"As you wish," said Goran and Sarrus in unison.

"Our people have lost their way," said Ar'Gallious. "But we will guide them back to the way of the Dragon, as it was in the beginning."

"And we'll make sure that no one stands in your way," said Goran.

"I know you will," said Ar'Gallious. "It's time to begin."

#

In the workshop, Tommy continued researching everything he could about Galaxy Tektronics, feeling he was close to finding a backdoor into the Mk17's security system. As he sat fixated on the computer, the sliding door of the communication room opened. Goran and Sarrus had returned.

"Well," said Tommy, swiveling in his chair to face them. "That only took forever. Thought you guys forgot about me."

Goran stepped forward.

"You are, as your people say, a lucky man, Lance," said Goran. "It appears our leader still has a use for you. In fact, Ar'Gallious wishes to speak with you now."

Surprised that Ar'Gallious still wanted to do business, Tommy never dreamed he would get a chance to address the leader of the Dragen Alliance. Until that moment, he had only ever dealt with the alliance through proxies such as Goran and Sarrus.

Ar'Gallious knew that if he were to raise an army powerful enough to overthrow the Lyrian Empire, he would have to forge alliances with outsiders. So, he was no stranger to working with unscrupulous types like Jax West and Tommy Lance. People like Tommy had no loyalties, but they were a necessary evil for Ar'Gallious to achieve his vision. To make matters worse, Tommy knew this and aimed to milk that relationship for all it was worth.

Now that he had the ear of the Dragen Alliance leader, Tommy felt closer than ever to achieving his own goal, becoming an exclusive shadow-market supplier to the Dragen alliance, a deal that would become more lucrative once Ar'Gallious overthrew the Lyrian Empire.

"Well, it's about time," said Tommy. "Lead the way."

While Goran escorted Tommy to the comm room, Sarrus moved toward the Mk17 storage unit sitting on the workbench.

He didn't know much about the Benton chamber, but if his leader was right, this advanced piece of technology, the answer to the galaxy's energy crisis, carried with it the potential to bring empires to their knees.

A short time later, Goran returned and stood by his brother's side.

"Well," said Goran. "It's time. You ready?"

Sarrus looked once more at the Mk17 container, then at his brother.

"Alright," said Sarrus. "For the glory of The Alliance."

#

Twenty miles off the coast of the beach leading to Tommy's safe house, Donald and the rest of Saber Team Seven secured a small, uninhabited island to serve as a staging ground for the forthcoming assault. The team made planetfall from orbit using a Marauder-class stealth dropship. Donald chose the Marauder because it doubled as a submersible, a necessary feature for his plans to conduct an amphibious landing on the nearby beach. With the island secure, Donald ordered Alex to call in the rest of the assault force.

"Saber Team One and Three will make planetfall soon," said Alex. "Long-range artillery support will be on station in thirty mikes."

"Outstanding," said Donald, anxious now more than ever to put an end to this wearying task.

Within minutes, Donald looked to the sky to see the retro-thrusters of two crescent-shaped Marauder dropships as they pierced through the dark thunderclouds swelling above. Moments later, the marauders touched down on the island's spacious beach, each filled with an eight-man shadow-tech team. As the soldiers spent the next half hour setting up the staging area, a massive four-legged artillery droid descended from the sky, landing on the island with a loud thud. As the slow-moving juggernaut moved into position, the ground shook as each leg of the metallic beast planted itself into the amber-colored sands below.

When the artillery unit reached its position, the massive droid lowered its broad body into the sand, after which dual turrets extended from atop its cube-shaped body. Once in position, the artillery team leader exited from the belly of the droid, approaching Donald with the hurried stride of a man ready to get down to business.

"Sergeant Jonah Parrish of Echo Team Six, reporting for duty, sir," said Jonah while saluting Donald. "It's an honor, Captain."

"Lookin' forward to working with you, Sergeant," said Donald while shaking Jonah's hand. "What's your status?"

"Long-range artillery is on station. Awaiting your command."

"Excellent. We'll be heading for the beachhead in fifteen mikes." Donald turned toward Alex. "Lieutenant Chavez, form 'em up."

"Fall in," shouted Alex, prompting the soldiers of Saber Team One, Three, and Seven to fall into formation in front of Donald.

The fully armored shadow-tech soldiers formed three squads of eight, with Saber Team Seven in the front row. Alex moved by Donald's side.

"At ease," called Alex, causing the twenty-four soldiers before him to move to parade rest.

"Listen up, people," said Donald. "Our objective is a portable Mk17 storage unit located somewhere on site." He transmitted the holographic likeness of the container to the heads-up display in each soldier's facemask. "The total number of hostiles inside is unknown. But expect heavy resistance."

"Rules of engagement, sir?" asked a soldier standing in the second row.

"You are weapons free on this one," said Donald, much to the delight of the combined Saber Team Squads. "If it moves, put a round in it."

"Yes, sir," yelled the soldiers in unison. With many of their operations being clandestine in nature, they were thrilled to do some real fighting for a change.

"We'll make landfall about three miles from the safe house. When we hit the beach, we move in tactical formations," said Alex, giving his last-minute instructions. "This is the real deal, boys and girls. Let's get it right," he concluded. "Fall out and fall into your Marauders."

After the soldiers left the formation, they filed into their respective landing craft, all of which were repositioned by their pilots to the water along the shoreline. Donald gave final orders to the artillery unit, then moved toward his team's Marauder, grabbing his rifle and equipment pack along the way. Though too far to see with the naked eye, he looked toward the direction of the target beach. *It's time to end this*, he thought while donning his own armor-plated facemask. He then took a deep breath to focus his mind, entered the craft, and sealed the hatch.

CHAPTER FORTY

From their new forest staging ground, Marcus finished laying out the infiltration plan to his team. A short time later, they resumed their half-mile trek in the direction of the safe house. As they worked their way toward the edge of the woods, leapfrogging between trunks of trees large enough to hide two adults, Marcus noticed something in the distance. He signaled for the team to halt. He lowered his tac-shades to get a better look.

Near the edge of the tree line leading to the target beach, Marcus picked up three thermal signatures in his tac-shades. It was Tommy's thugs, two of whom were getting high on zeth, while the other stumbled drunkenly toward an overgrown bush to relieve himself. With their rifles strewn about the ground like discarded toys, the guards laughed and joked with each other, oblivious to their surroundings.

Using a series of hand gestures, Marcus signaled to his team he had eyes on three targets, and that he wanted them taken out quietly. Skye and Tony moved forward to neutralize the mercs while Marcus, Daren, Jason and Jack spread out amongst the trees to cover them. As the zeth-inhaling thugs took another hit, Tony overtook them before they realized he was there. He grabbed the two mercs by their throats. With massive cybernetic hands crushing their windpipes, the thugs stared at Tony in horror, seeing only the golden glow of their attacker's cold pupil-less eyes.

Tony slowly lifted the mercs from the ground, as they groaned, kicked and squirmed to no avail. While Tony hanged the two thugs with his bare hands, Skye approached the third just as he finished urinating into the bushes. As soon as the merc turned around, he froze when he saw the large cyborg choking out his two companions. The merc snapped out of it and reached for the sidearm strapped to his leg, but before he could touch the weapon's grip, Skye unsheathed two razor-sharp combat knives with curved, twelve-inch blades and attacked.

With two swift hooks, Skye severed the left and right jugular veins of the merc. He stumbled back, gripping the sides of his neck to keep the blood from exiting his body. But the effort was futile. The guard dropped to his knees, then flat on his face where he bled out on the teal-colored soil of the forest floor. The team knew these mercenary types all too well—heartless savages and killers. And Marcus figured the galaxy would be better off without them. So, he gave the order to kill any mercenary they encountered without hesitation. As Tony dropped the lifeless bodies of the two mercs to the ground, Marcus and the rest of his team moved from behind their respective trees to join Skye and Tony.

"Alright, let's do this," said Marcus to the crew. "And remember, no one makes a move on the safe house until I give the word. Got it?"

The team acknowledged and headed toward the edge of the forest, taking up positions behind the trees. With the entire beach littered with massive crystalline formations, Marcus figured the obstructions would make the perfect cover, so he ordered the team to stick close to them while making their way down the beach. To keep them from showing up on enemy sensors, Skye activated an electronic chip she had installed on each team member's belt. The device emitted a localized scrambling field, rendering them invisible to sensors. But time was of the essence, as the sensor scrambler had a limited power supply.

Within twenty minutes, Marcus' team moved within four hundred yards of the safe house, taking up positions behind a broad crystal formation. The rear of the safe house was swarming with at least fifteen mercs, though none of them were on high alert.

"On my mark," said Marcus softly to his team.

The team silently acknowledged. While Tony and Daren stood ready, Skye and Jason exchanged reassuring looks, ready to make the final push to the safe house together. Daren looked toward Jack. He looked fidgety. They locked eyes, and he didn't like what stared back.

Daren considered himself to be a functional addict, never taking more zeth than he needed to get by. However, he stayed away from *boosters,* the powerful, reflex-enhancing cocktail of drugs on which many mercenaries were hooked. And one look at Jack, and Daren knew that he had just injected himself with a massive dose of boost, hoping it would give him an extra edge in battle. But given his already volatile personality, Daren had a feeling that Jack was on the verge of a mental break.

Before Daren could tell the overactive merc to settle down, Jack rushed the rear of the compound well ahead of Marcus' signal. The entire team watched in disbelief as Jack ran toward the safe house, yelling like a madman, firing his rifle toward the startled guards in the distance, forcing them to scramble for cover. A few steps later, circular-shaped mines rose from the sands like mini flying saucers, with spikes rising from atop the floating discs. The mines flew toward Jack like a swarm of hornets, impaling his legs and arms, bringing him to his knees. Jack looked back toward Marcus and the team just before the rapidly beeping mines went off, blowing him to oblivion.

A weak electromagnetic pulse caused by the explosions hit the team, disrupting their scrambling fields long enough for them to trigger every proximity alert in the area. The entire team showed up on every mercenary's sensor grid embedded in their wristbands. Within seconds, the mercs aimed their rifles toward the crystal formation concealing the team and opened fire.

"Damn it," yelled Marcus, knowing that retreat was now clearly out of the question. "Open fire."

The team popped their heads up from behind cover and returned fire, immediately dropping several of their attackers. The surprisingly dense crystals absorbed the laser bolts, but it was no telling how long the structure would hold. They needed to press forward, but it was obvious from Jack's monumental blunder, that a massive field of smart mines stood between them and the safe house.

"Skye," yelled Marcus amid the chaos of the firefight. "Handle the mines. We'll cover you."

Expecting smart mines on every mission, Skye always brought the device that Max designed years ago that she dubbed the *rumbler*, but never used it before now. The device resembled an ancient claymore mine with two thick prongs extending from the bottom. The rumbler had two modes, one for slowly scanning and disabling smart mines, a feat that would take too much time given their current situation, and a secondary mode for which the device was named, used to detonate the concealed traps instantly. She low-crawled forward a few feet from the side of the crystal formation while blue-colored bolts of energy whizzed above her head.

Marcus and the rest of the crew poured on their covering fire, buying Skye enough time to jam the prongs of the rumbler into the ground. She quickly crawled back to cover, nearly catching several rounds in the process. Skye accessed the device from her wrist pad.

"Sending subsurface pulse," warned Skye to the team, prompting all of them to take cover. "FIRE IN THE HOLE!"

With a press of the button, a deep, loud sonic shock wave rumbled through the ground. A spectacular chain of explosions went off as every mine embedded in the ground at the rear of the compound went off one by one. But as was typical with many of the over-engineered devices conceived by Max, the shock wave that emanated from that small device, rippled and grew like the waves on a pond caused by a massive falling rock.

The shock wave raced from the rear of the compound, all the way to the front, rattling everything in its path like a low-magnitude earthquake, causing mines to continue to detonate all around the compound. The quake was so strong that it knocked everyone on the battlefield to the ground.

"Clear," yelled Skye, as she stood to her feet.

Marcus and the rest of the team picked themselves up from the ground and looked downrange toward the stunned guards in the distance.

"Let's go," yelled Marcus.

And with that, Marcus and his crew rushed the rear of the compound.

CHAPTER FORTY-ONE

Donald and his team arrived on foot just as the sonic pulse that originated from the rear of the compound reached the front. The smart mines had an internal feature that gave them the ability to distinguish between friendly and unfriendly targets. However, the rumbler's powerful shockwave disabled that critical safety feature, causing the mines to detonate irrespective of who was standing near. The blast caught many of the unsuspecting mercenaries well before any of them could respond to the firefight taking place in the rear of the compound.

"What the hell?" said Donald as he witnessed explosion after explosion ripping through the unsuspecting guards. Deciding to seize on the opportunity, Donald turned to Alex. "Call it in, now."

"Echo Six, Saber Seven," yelled Alex into his comm unit. "Adjust fire, over."

"Saber Seven, Echo Six, adjusting fire," responded Jonah from the small island twenty miles away as he eagerly awaited firing coordinates from Saber team.

"Grid: 348—752. We have tangos in the open! Direction: 2300, range 2-5-0-1 meters. Fire when ready, over," said Alex into the radio as he and the rest of Saber team readied themselves for the coming storm.

"Roger that, firing on grid," replied Jonah. "Sendin' the rain, over."

Within moments, a thunderous boom resounded across the land as the powerful artillery cannons of Echo team went off, hurling its high-yield explosive rounds toward the guards near the safe house. Those fortunate enough to have survived the exploding mines were blown apart by artillery rounds that hit their mark with deadly precision.

Donald ordered his team to converge on the safe house. The twenty-four soldiers of Saber Teams One, Three and Seven rushed the rear of the compound using the boulders and crystalline structures as cover, gunning down the remaining guards left standing. But as they closed the gap to the safe house, automated turrets on the rooftop arose from their hidden compartments and opened fire, hurling a blur of golden-colored energy bolts toward the team. The rounds killed and maimed many of the soldiers from Saber Teams One and Three, forcing the rest into cover.

"Take out those damned turrets," yelled Donald to Alex. "But I need the building intact." He watched in dismay as the menacing cannons above continued to rain fire on the soldiers under his command.

"Echo Six, Saber Seven! I have sentry guns on the roof," yelled Alex to the artillery team over the radio. "I need airburst rounds delivered downrange. Adjust fire as follows."

Alex called in new firing coordinates, prompting the artillery team to send a fresh volley of airburst rounds toward the rooftop turrets. The rounds detonated above the targets, sending a directed electromagnetic pulse that overloaded the cannons, causing them to shut down, while keeping the building relatively unscathed.

"Echo Six, Saber Seven. Cease fire, cease fire. Good shot," yelled Alex into the radio.

Donald signaled the go-ahead for the remaining soldiers to rush the compound, just as reinforcements flooded the area from inside the building. But the Shadow Tech soldiers of Saber team were undaunted, meeting the opposition head on with great zeal and ferocity. They eliminated the guards with extreme prejudice, forcing the survivors to cower back inside the compound, sealing the door behind them.

Upon reaching the sealed doors, Donald turned to staff Sergeant Paul Edwards, team leader of Saber Team One. He could tell Paul struggled with the loss of so many close friends. However, Paul, like the surviving soldiers of all three Saber team units, did a remarkable job of keeping his emotions in check. Their steadfast resolve amid hardship was a testament to the extraordinary discipline of all Shadow Tech soldiers.

"Sergeant Edwards. I want you to take command of Saber Teams One and Three. I need you to evac the injured and the dead to the Marauders. Lieutenant Chavez and I will take Saber Team Seven to retrieve the package. Once we have it, we'll rally back on your position."

Paul acknowledged with a salute. He gathered the soldiers under his command, then hurried off to carry out Donald's orders.

"Alright, Lieutenant Chavez. Let's go in," said Donald.

Alex signaled for one of their heavy weapon specialists to breach the door. He lined the door with a grayish liquid explosive that was foam-like in appearance. He moved back, then signaled for the team to take cover.

Seconds later, the explosive blew the door off its hinges, sending it flying into the building, killing several guards. As the smoke cleared, Donald and his team stormed the compound, meeting the over-matched guards head on without fear and without mercy.

#

Guided by Daren's tracking device, Marcus and his team negotiated the maze of hallways in the building's rear, meeting little resistance along the way. They could hear the fighting going on in the compound's front but didn't know what was going on. But in that moment, Marcus didn't care. Anything to take the heat off them was a fortunate turn of events from his perspective. And given the sheer amount of guards Tommy had on site, Marcus figured they needed all the breaks they could get.

"C'mon, not now," said Daren while banging his hand against the side of the small rectangular-shaped tracking device. "The signal disappeared."

"Where is it?" Asked Marcus.

"Last hit came from that room," said Daren, pointing toward a room just down the hall from their current position.

#

In the front of the safe house, Donald and his crew finished mopping up the last of Tommy's thugs.

"Where's the container?" asked Donald while shoving a fresh energy cartridge into his rifle.

"The device is no longer showing up on the scanner. But the last confirmed signal came from that room up ahead," said Alex, pointing to the solid metal doors leading to Tommy's workshop.

"Sounds good. Let's go," said Donald.

But before they could converge on the workshop, a fresh contingent of guards ambushed them, spilling into the room from doors that lined the hallway. The new volley of shots killed one of Donald's men, prompting the rest of the team to take cover and return fire. Furious and ready for payback, Donald and his team pressed their attack, ready to put an end to the opposition once and for all.

#

In the rear of the complex, Marcus and his team reached the solid metal door leading to the workshop. Skye tried to bypass the door's access code, but the security measures were so advanced, her scything device would have taken ages to decipher its code.

"It's a no-go, Marcus," said Skye as she detached her patch cable from the control panel.

But Marcus hadn't traveled this far to be stopped by a single door. So, he signaled for Tony.

"Step back," said Tony as he approached the door.

Like everyone else, the mission to Protos IV irritated him. It was time to bring an end to this burdensome ordeal. Using his powerful cybernetic leg, Tony channeled all his frustration and anger into one kick, knocking the entire door off its hinges, causing it to fall to the floor with a loud thud. They rushed the room, exchanging fire with Tommy and three of his thugs. But Marcus, Jason and Tony landed the fatal shots that eliminated the three mercs. Seeing that he was hopelessly outnumbered, Tommy tossed his gun to the floor, holding up his hands in surrender.

"Okay, I give up," said Tommy.

Once inside the workshop, the sound of the battle raging outside the front door was louder than ever.

"I don't know what's goin' on out there. And we ain't stickin' around to find out, so hurry this up," said Marcus to Daren.

Daren holstered his sidearm and approached Tommy with a hurried stride. And before Tommy could open his mouth to utter some clever, ill-timed quip, Daren floored him with one punch.

"Where is it?" shouted Daren, struggling to keep himself from pulling his sidearm and blowing Tommy's head off.

"Daren, buddy," said Tommy after spitting out blood. "You're alive. Hey, no hard feelings, right?"

But Daren was in no mood for games. He grabbed Tommy by the collar and snatched the shades from his face, throwing them to the ground. He pummeled his former business partner until his fists bled. When his hands were too sore to punch, he shoved Tommy to the ground and stomped him into the metal floor.

"Oh, you think I'm playin' with you?" said Daren as he unholstered his XP-90. He ordered Tony to seal the front door and told Jason and Skye to tear the workshop apart in search of the Mk17 container.

Daren turned his attention back to Tommy, grabbing him by the collar. He shoved the barrel of his handgun hard into Tommy's forehead. "Start talkin' and I swear I'll make it quick."

"It's gone, man. They took it," yelled Tommy.

Daren felt as if someone punched him in the gut. He released Tommy and took a few steps back. "What?" yelled Daren.

Marcus stepped forward. "Who took it?" he asked.

"Well, where do I start?" said Tommy as he crawled to retrieve his shades from the ground, strapping them back on his face. He staggered to his feet. "Oh yeah, that would be those two backstabbing rat-bastards, Goran and Sarrus."

"Your Lyrian buddies?" said Daren, laughing at the sheer irony of it all.

"That would be them," said Tommy, sighing in embarrassment. "But tell me, Daren, since when did you start workin' with The Guard?"

"What are you talking about?" said Daren, genuinely confused.

"So, you're not with those over-caffeinated commandos out front, tearing my place apart?" said Tommy.

"ISG? We need to move. Now," said Marcus.

"Best idea I heard all night," said Tommy. "I tagged the box, so we need to go before we lose that signal."

"Hold on now, *we* ain't goin' anywhere," said Daren as he aimed his handgun at Tommy.

"Come on, is that how you'd do your old partner?" said Tommy. He turned to Marcus. "But I have to admit, I'm surprised the two of you are still working together, you know, after what Daren did to you."

Marcus wasn't sure why Tommy's words resonated with him. At first glance, he took Tommy to be a guy that would say anything to save his neck. But still he felt compelled to know more. But he had to intervene, as Daren appeared to be moments away from putting a round between Tommy's eyes.

"What's he talkin' about?" asked Marcus, as he knocked Daren's hand away, causing him to fire a shot that narrowly missed Tommy's head.

It surprised Daren that Marcus had intervened on Tommy's behalf.

"Oh, he didn't tell you?" said Tommy, shaken from having just seen his entire life flash before his eyes.

Tommy briefly recounted the Titan job eleven years earlier and how it was he and Daren who had planned and executed the side job that caused Raven Squad's primary mission to go south, resulting in Marcus' arrest.

"That's why he went after those crates," continued Tommy, watching the look on Marcus' face. "But I'm sure he didn't mean to get you busted."

Marcus was livid. He always knew Daren to be reckless and impulsive, often making rash decisions at the spur of the moment. But he couldn't believe that Daren would intentionally do something so risky without telling him first.

"You mean to tell me..." said Marcus, unable to complete his own sentence. The thought of what Daren had done made his stomach turn.

As angry as he had been all those years, Marcus never believed Daren's actions on Titan to be more than a bad call, made at the worst possible time. But one look at the shame in Daren's eyes, and Marcus knew that Tommy was telling the truth. Enraged, Marcus felt detached from himself, having no control over his body. He forgot all about the ISG soldiers fighting outside that front door who could literally bust in at any moment. And before Daren could even open his mouth to explain, Marcus tackled him to the floor, locking his hands tightly around Daren's throat. Skye and Jason tried in vain to separate the two.

"Come on Marcus," pleaded Skye. "We have to get out of here."

Marcus couldn't hear anything beyond the pounding of his own heart as he tried his best to end Daren's life on that workshop floor, just as Daren had ended his back on Titan.

Tony finished sealing the front door to the workshop, hoping to slow down the ISG soldiers fighting outside. He turned to see the chaos going on behind him. But before Tony could intervene to keep Marcus from killing Daren, the door he just sealed was blown apart. Shadow Tech soldiers clad in black stormed the room, opening up on Marcus and his team. As the heat of a crimson-colored energy bolt grazed the side of his cheek, Marcus snapped out of his fit of rage, rolling in to cover.

Tony rushed forward to drag Daren out of the line of fire, catching a round to the shoulder in the process. The rest of the crew exchanged fire with Saber Team, but it was clear from the start they were hopelessly outgunned.

"I got the last of my guys closing in on the compound," yelled Tommy to Marcus. "When they get here, we can slip out the back."

Marcus nodded, ordering his team to hold their ground.

Amid the chaos, Donald saw something from across the room, forcing him to remove his helmet to get a better look.

For a moment, Donald and Marcus locked eyes.

"Marcus?" said Donald to himself, certain his eyes were playing tricks on him.

From across the way, it also shocked Marcus to see a face he thought he'd never see again.

"You've gotta be kidding me," said Marcus as the last of Tommy's guards flooded the room from the rear exit. Marcus glared at Daren. "This ain't over," he said. He turned toward the team. "Fall back to the ship, now."

Daren turned to Tommy.

"You're coming with us," said Daren. He struggled to speak after being nearly choked to death. "For now."

A new batch of mercenaries stormed the room, forcing Donald and his team to redirect their fire to the greater threat coming from the rear. As they did so, Donald noticed Marcus and his crew making a mad dash for the rear doors.

"Oh, you can run..." Said Donald, as he took out three more mercenaries.

While he didn't have the resources to give chase, Donald vowed to once again track Marcus down. But this time, he had no plans on leaving Marcus' fate to the judges of the Interstellar League justice system.

CHAPTER FORTY-TWO

Having barely escaped Protos IV with their lives, Marcus and his crew were back aboard *The Indicator* with their new prisoner, Tommy Lance. It had been less than two hours since their clash with Donald and his team on the planet's surface, so the crew wanted nothing more than to put as much distance between them and that planet as possible.

As the vessel darted silently through hyperspace, Jason sat behind the ship's helm controls, grateful for the opportunity to once again pilot the ship that he so greatly loved. At first, he was sure they would have been blown from the stars the moment they left the planet's atmosphere. But to Jason's great surprise—and relief—he found the stealth technology that he and Tony installed on the ship years ago to be present and fully operational. Were those covert systems not present, they would have never made it on or off the surface of Protos IV without ISG forces detecting them.

But it wasn't all smooth sailings for the former members of Raven Squad, as the ship's weapons that were still on board had been totally disabled and partially dismantled. A necessary modification, or Marcus would have never been allowed to dock the ship back at his home star port.

Jason was irritated with Daren for never bothering to tell Marcus the truth about why they were traveling to Protos IV in the first place. If he had, they would have probably had time to install *The Indicator's* primary armament, which would have made his job much easier. But there was nothing Jason could do about it, so he continued to do the best he could with what he had.

After jumping to and from several star systems, Jason was finally satisfied that they were out of the danger zone. He then placed the ship into a standard circular orbit of a nearby Gas Giant and rejoined the rest of the crew in the cargo bay. When Jason arrived, everyone was in a heated discussion regarding recent events on Protos IV, with Tommy clearly in the hot seat.

"C'mon, guys, I told you everything," said Tommy amid a sigh of frustration.

"So, in essence, all you really have is an unreliable tracking device and a hunch?" responded Daren after listening to Tommy's half-baked plan on how to retrieve the Benton chamber from the clutches of Goran and Sarrus.

"Well, when you put it that way," said Tommy, "I can certainly see your concern..."

Daren leaned against a nearby cargo container and pulled his XP-90 from its holster.

"Remind me again why I didn't kill you back there," said Daren.

As much as he wanted to shoot Tommy and eject his body out of the nearest jettison chamber, he still needed him if there was to be any hope of tracking down Goran and Sarrus. So, he holstered his sidearm and approached Tommy.

"Your crappy long-range tracker can *maybe* tell us the system they're in, if that. What if you're wrong about exactly where they're headed?" said Daren.

"Look, man, I've been doin' this a long time. It's no hunch," said Tommy. "I'm tellin' you, there's no way they can open that Mark17 without that egghead programmer, Brad Jennings. That's who they're goin' after. I know it."

"Right," said Daren. "The twenty-seven-year-old, zeth-addicted hacker who was ousted from Galaxy Tektronics two years ago? C'mon, man, are you serious?"

"Think about it, Daren. Galaxy Tek stood to make trillions off those portable Mk17s," said Tommy, recounting the hours he poured into researching the little-known incident that had occurred at one of the galaxy's largest cargo container manufacturers. "They weren't gonna let some random code-jockey sound the alarm about a little-known backdoor into their so-called 'impenetrable' security software."

"It does kinda make sense," said Skye, disgusted that she even agreed with Tommy on his point. "I mean, Brad Jennings is a legend among scythers. He's supposedly the only guy to crack a Mark17. And believe me, that is no small feat. Man, I wish I could've seen his algorithms."

"Finally, a voice of reason," said Tommy, relieved that someone was starting to take him seriously. He turned to Skye, flashing that signature grin. "You know, I have something way more interesting than *algorithms* I can—"

"Um, I'll pass on that," said Skye, killing the thought before Tommy could finish thinking it. "But here's the problem," she continued. "Brad went off the grid after the Galaxy Tek thing. He hasn't been seen since."

"Like I said. They're headed to Xenon Five... Brad's last known location. That's where I'd start," said Tommy. "All we have to do is ambush Goran and Sarrus before or after they make their move on our little junkie scyther."

"This is a stretch, Tommy," said Daren. But he was prepared to entertain anything at that point, as they had no other way of catching up with the two Lyrians.

"How did you come across this information in the first place?" asked Tony, skeptical of every word spewing from Tommy's lips. "A security breach that significant would've been all over the news."

"We're talkin' about a company with a lot of money, my cyborg friend. Trust me, they'd have no problem keeping something like that quiet," said Tommy as he stood from the cargo container upon which he had been sitting.

"But to answer your question," Tommy continued, deciding to come clean to everyone. "Once Goran left the room to talk to his boss, he left his computer terminal unlocked. Somehow, he managed to break into Galaxy Tek's corporate databanks. I mean, the guy had no idea what he was on to. So, I *furthered* the research while he was away," he continued with a self-absorbed look on his face. "And that's when I stumbled on Brad's file. Apparently, his former employer is lookin' for him too. That's how I found his last known location," said Tommy, proud of his extensive sleuthing efforts.

Having heard everything he needed; Jason stepped forward.

"Maybe he wanted you to think that's where they're headed, you idiot," said Jason. "I'm pretty sure they knew we'd be hunting you down after that crap you pulled on Argos. Hell, they're probably expecting you to lead Daren straight to them so they can beat the codes to the Mk17 out of him."

Though he often didn't act like it, Jason was highly educated, a man of both reason and logic, the result of the extensive schooling he received on Earth while living with his father. Jason had the uncanny knack for quickly making connections that others couldn't see. Skye often said he was too smart for his own good, always over-analyzing situations. But this time she, like everyone else, was in complete agreement with his assessment.

"Trust me," said Tommy, dismissing Jason's comments. "I know these guys. They ain't that bright."

"Oh, but they're bright enough to break into Galaxy Tek's databanks, right?" countered Jason.

"Why don't you run back to the bridge and play with your little *flight stick*, kid," barked Tommy, moving face to face with Jason, "and leave the strategy to the adults in the room?"

Jason shoved Tommy hard, nearly knocking him off his feet.

Daren stepped in to keep a brawl from breaking out in the cargo bay. But before Daren could open his mouth to weigh in, Marcus stepped forward after having sat quietly in the corner during the entire conversation.

"Well, this is all fine and good. But this is where we part ways. I'm done," said Marcus.

"Come on, Dek," said Daren, trying to keep a bad situation from becoming worse. "I need you on this one."

"You need me?" snapped Marcus. He moved face to face with Daren. "I've done way more than I should've here. Not only could I end up back in prison, but I put my family at risk foolin' with you."

Daren started to speak, but Marcus immediately cut him off.

"I'm droppin' y'all off at the nearest star port," said Marcus to the rest of the team. He turned back to Daren. "And as for you, I never want to see your face again."

Marcus turned to leave the area, looking back just before exiting the cargo bay. He and Skye momentarily locked eyes. She started to approach, but Marcus turned and exited through the bulkhead door, closing it behind him. Once inside the narrow corridor, Marcus banged his fist against the unforgiving metal wall. The sound reverberated throughout the empty walkway. He took a deep breath to compose himself. He was done with his old team, and with his old life.

In an instant, all the lingering frustration he felt since his release from prison was suddenly gone. All he wanted in that moment was to be there for his family, even if it meant working a mundane, low-paying job for the rest of his days. But Marcus was angry with himself, because he knew that Daren was up to no good the moment he showed his face on Stratus One. But he took the job anyway, which turned out to be the biggest mistake of his life. But no matter how Marcus tried to convince himself otherwise, he took that job because what he had wasn't good enough.

As much as he loved his family, the straight and narrow was a path far too difficult to travel, often causing him to feel as if he were suffocating, in desperate need to simply breathe again. Despite his anger, he wished there was a way to convince Skye to come with him so they could start anew. But he knew that would never happen, so he never bothered to ask. As he strode the cold steel deck of the ship toward the bridge, he couldn't help but wonder if his recent change of heart had come too late. Frustrated and fearful that he had finally crossed a line from which there was no return, he entered the bridge of *The Indicator*, wondering how he could ever face his family back home.

#

Less than an hour later, *The Indicator* jumped into the Kronos system. Kronos was a small red-dwarf star with only one habitable planet in the entire system. The residents of that vibrant world called the planet *Helios*. Marcus navigated *The Indicator* toward Aris, the large star port proudly orbiting Helios and requested permission to dock. Within moments, he received clearance and began the landing sequence.

The scale of the hexagonal-shaped star port was impressive, though it paled in comparison to Stratus One. But Aris was still the largest trading outpost in the sector, so Marcus figured that Daren and his crew would have little trouble securing passage on another starship. Marcus landed *The Indicator* on the circular landing platform of Aris' docking bay. After cutting the engines, Marcus lowered the aft loading ramp, exited the bridge, and silently ushered the crew from his ship, never making eye contact with any of them.

Daren and his companions unceremoniously exited the ship, joining the hordes of passengers on the star port as they scurried to and from enormous transports. He looked at the numerous starships perched upon the forty-seven docking platforms lining the gargantuan docking bay and sighed. He then looked up from amongst the crowd toward Marcus, who stood quietly near *The Indicator*'s aft loading ramp. Not knowing what to say, Daren kept quiet, holding out hope that Marcus would somehow have a miraculous change of heart. But there would be no miracles that day.

As Marcus stood atop the loading ramp, he looked over to Skye. He knew there was no way she would come with him; and given everything that had transpired, there was no way he could stay. But the look Skye gave him before shifting away her gaze pierced his soul, causing him to fight back the foreign sensation of tears. But he steeled himself and hit the button to his left, causing the ramp to retract into the ship.

Daren's heart sank as he watched Marcus turn his back to them just before the loading hatch sealed shut with a loud clank. Within minutes, *The Indicator*'s engines roared to life, causing a slight rumble through the deck of the star port. The ship gently lifted from the landing pad as its maneuvering thrusters gracefully propelled it toward the shielded exit bay in the distance. Daren watched *The Indicator* until it finally left the star port, leaving him to stare at the exit bay as its massive hanger doors closed. Moments later, Tony approached.

"What now, D?" asked Tony, looking to his longtime friend for direction.

"We secure another transport," said Daren, determined to finish what they started.

Tony put his massive hand on Daren's shoulder as Skye and Jason moved by his side.

"We're with you, brother," said Tony, with Jason and Skye echoing his sentiment.

"Alright, people," said Daren after taking a deep breath, "let's move."

CHAPTER FORTY-THREE

Teric Winters stood gazing through the transparent metal wall of his lavishly decorated grand hall. It was a massive chamber filled with exotic, high-tech gadgetry from across the galaxy. The view from his tower overlooked the picturesque landscape of the planet known to outsiders by its stellar designation, Beta-Three-Four-seven-nine.

A largely rocky planet located in the farthest reaches of the outer core, Beta-Three-Four-seven-nine was thought to be inhospitable to life. In fact, the scientific community of the inner core star systems believed the celestial body was ravaged day and night by powerful storms, intense seismic events, and extreme volcanic activity. Therefore, the planet's original discoverers, the Erecians, logged the planet as Class Zero, indicating that it was beyond the Galactic Planetary Consortium's ability to colonize. So, they assigned the planet its stellar designation for the astral archives, then moved on to the next system.

From Teric's point of view, Beta-Three-Four-seven-nine was anything but inhospitable. From his perspective, that wondrous world represented the very salvation of the beloved movement founded by his late father, Zedolph Winters, hundreds of years earlier, the movement now branded a terrorist organization by the rest of mankind, Orion's Shield. As he stood alone in his chamber awaiting an overdue update from Daren West, Teric marveled over the impressive landscape before him, rubbing his graying, well-groomed beard with his fingertips, a habit he often had during times of deep reflection and contemplation.

Teric turned to fix his gaze on the twin mountains standing proudly in the distance, between which he could see the silhouetted profile of a gargantuan sensor-scrambling array, resembling a cluster of oversized satellite dishes and solar panels. Believed to have been left behind by the same hyper-advanced race that once inhabited Mars eons ago, the scrambling arrays, scattered throughout every continent on the planet, greatly exaggerated the surface conditions on the world with false readings so intricate they could fool even the most advanced orbital scanners, helping to conceal all presence of life on the planet's surface.

During the time of the Great Purge, Teric sent hordes of his finest scientific minds to study the massive arrays after they were first discovered by survivors of an Orion's Shield capital ship that had crash-landed on the planet. *It was destiny*, Teric thought. He later moved the base of operations for Orion's Shield to Beta-Three-Four-seven-nine, vowing to finish the work of his late father, ushering in a new era for humanity, shaping the destiny of man according to his own vision.

Teric went on to name that remarkable world after the Erecian word for *destiny*. Teric called the planet, *Zirakuza*, a world known only to the most radical and loyal members of Orion's Shield. And now, hundreds of years after first discovering the scrambling arrays on Zirakuza, Teric was amazed its mysterious power source never once fluctuated. *If only we could replicate this raw power, there'd be no need for the Benton Chamber. And I wouldn't have to deal with such staggering levels of incompetence*, Teric thought, referring to Daren, whose grainy holographic transmission finally came through. He returned to his antique desk at the center of the grand hall to receive the distorted holo-trans.

"Where is my chamber, West?" snarled Teric.

"The Dragen Alliance rats that Tommy was workin' for made off with it," Daren replied nervously, dragging a handcuffed and gagged Tommy into view of the holo-camera for Teric to see. "The ISG showed up too. We barely got out, but Tommy says he knows where the two Lyrians are headed."

"Is that so?" said Teric, casting a stern glare at Tommy. "Make no mistake, Lance," Teric continued, his cold eyes meeting Tommy's, "you will die for this treachery." He rose from his chair. "But how you die depends on whether or not you reclaim my prize." He walked toward the transparent wall in his chamber, turning his back to the holographic images of Daren and Tommy. "But know this... If you return empty handed, I'll make you watch as I personally cleave the flesh from your bones."

Daren shoved Tommy away from the holo-cam. He started to go over his plan to retrieve the chamber from Goran and Sarrus, but Teric interjected.

"Enough," said Teric, never turning his gaze from the serene Zirakuza landscape. He grew tired of Daren's voice. Tired of the excuses. Tired of a destiny unfulfilled.

"*You* are on borrowed time, Mr. West," said Teric. "And that time is running out," he said, finally turning to face Daren. He could almost see Daren's heart pounding through his chest. "I don't care what you have to do. I want that chamber in my hands by the end of the week." With that, he terminated the communication.

Teric had lived for many centuries. His unnaturally long lifespan was the result of countless feats of banned biological engineering. Yet in all his years, he had never been more furious than he was in that moment. As Teric stood watching the trees in the distance, swaying violently in the strong Zirakuza winds, he closed his eyes and sighed, knowing that if he didn't recover the Benton Chamber soon, the ones who charged him with that task would be most displeased. He shuttered at the very thought. There was very little that frightened Teric beyond the prospect of dying. But his silent benefactors were a vicious race to be both feared and revered. Failure was simply not an option.

As Teric returned to his desk to activate his levitating comm unit, he fell to the ground. His head pounded, feeling as if his brain were being split in two. But as suddenly as the pain overtook him, it stopped. Rubbing his temples, Teric staggered to his feet. But before he could take another step, the disturbing sensation returned, forcing him to his knees. The pain intensified, leaving him helpless, lying on the ground in a fetal position. It was as if a thousand voices were invading his mind, screaming in pitches both high and low, yet speaking with a singular voice. Moments later, the pain and the noise subsided, allowing him to crawl to his desk to activate his comm device.

Samson Kull's deep, raspy voice responded on the other end.

"Come to my chamber immediately, my old friend," said Teric, panting as if he had just run a marathon. This wasn't the first time the voices bombarded his mind, speaking to him in their unnerving shrill. And he knew exactly what it meant. "I fear the situation has... escalated."

CHAPTER FORTY-FOUR

Within minutes, Samson stormed into Teric's grand hall to find the leader of Orion's Shield motionless in the chair behind his desk, eyes wide open as if he were seeing images no one else could see.

"Teric," Samson yelled, trying to snap his friend and leader out of his unresponsive state.

Samson checked his vitals. To his relief, Teric had a pulse and was still breathing. He called his name again, but there was no response. He examined Teric's eyes. His normally dark pupils were now gray and fluttering. He started to call the medics, but Samson stopped himself. He knew exactly what was happening. And no one else could know of this dark secret. He had to let the moment pass. But given Teric's current condition, Samson was unsure if his friend would even survive this latest encounter.

#

Teric opened his eyes to find himself in a room ten times the size of his own grand hall. But there were no treasures or lavish décor to be found. No awe-inspiring displays of wealth and excess like that which permeated Teric's grand hall. He saw only a black room with stone floor tiling, upon which symbols of an unknown language were intricately carved. Some of the symbols on the floor emitted a faint luminous mist ascending like a glowing fog, barely rising above his ankles. Teric looked up to see sharp stalactites jutting from the ceiling like the tips of massive spears, threatening to impale the unwary lurking below. The room was cold, and the air thin, carrying with it a putrid stench, like that of decaying flesh.

Teric inched his way forward, disoriented and struggling to breathe. He tried yelling for help, but no sound could escape his mouth. Despite the bitter cold, Teric began to sweat profusely as the uncharacteristic sensation of fear gripped every fiber of his being. As he slowly traversed the largely empty room, he looked about his surroundings, seeing walls aligned with a series of engraved torches. But instead of fire, each torch contained dark crystals that emitted a purplish-black glow, the only source of light in the room.

As Teric strode the stone floor, the temperature in the room dropped sharply, causing him to shiver uncontrollably. Suddenly, the floor began to vibrate. An eerie dark mist formed in front of him. From the shadowy mist, an ominous throne of dark metal and stone appeared. Upon the throne sat a thin, shadowy figure slouched in his chair, gripping a multi-bladed spear in his left hand. Only the figure's pupil-less eyes, burning bright with a purplish-black glow, peered from its dark, featureless face.

Every bone in Teric's body seized with fear, having never seen such a sight in all his centuries of existence. And despite its darkened visage, he knew exactly who the shadowy figure was, though he had never seen him in this form. Instinctively, he dropped to his hands and knees, staring at the floor as if bowing before a king.

"Lord Kalen," said Teric. He tried to speak once more, but the words could not escape his lips.

"What was our agreement, Mr. Winters?" said Kalen, his boisterous voice booming throughout the throne room.

Then, as if by Lord Kalen's will, Teric's voice returned to him.

"That I would deliver the Benton chamber to your hands," said Teric in a slow, shaky voice. "And in return—"

"We would send our armadas before you, that you might bathe in the blood of all who oppose your will," shouted Kalen, finishing Teric's statement. As he spoke, the ground shook as if they were at the epicenter of a great earthquake. But as his voice softened, the tremors subsided. "We promised you power, Mr. Winters," he continued, his deep and distorted voice sending chills down Teric's spine. "All I asked was that you deliver the device that the infidels use to defile our people's sacred crystals."

Fearing that death was at hand, Teric found the courage to look upon the throne.

"No," yelled Teric. "I won't fail you." But when he looked up, the chair was empty. He stood to his feet, looking around. "But I need help," he yelled to the empty room, certain the shadowy figure could hear him. "You see what I'm forced to work with."

As Teric spoke, Kalen materialized in a dark mist behind him. Teric turned to find himself staring at the glowing blade of Kalen's twisted spear. Imbedded in Kalen's metal gauntlets were the same crystals resting atop the torches throughout the room. The crystals seemed to pulsate as Kalen moved, fueling his almost supernatural powers.

Teric tried to speak, but it was like Kalen ripped the very words from his mouth. Teric again dropped to his knees.

"Perhaps our faith in you was, misplaced," said Kalen.

Dread overtook him. For Teric's greatest fear was not death itself, but dying with a destiny unfulfilled, like his father before him. He closed his eyes and exhaled, expecting to have his head mounted upon the spike in his tormentor's hand, where it would adorn the halls of the bitterly cold throne room. But the killing stroke never came. Instead, Teric opened his eyes to find the dark figure gone. He turned to find Kalen back on his throne, with eyes burning bright like a purple flame.

"But you are correct, Mr. Winters," said Kalen. "If this crucial task is to be fulfilled, you will need help," he said, leaning forward in his chair. "There is too much at stake. So, I will send a Goro'Sai."

"Wait, your exalted one," said Teric, suddenly finding himself able to speak. "A Goro-what?" he asked, confused. "And only one? Milord, if I could have just a few of your mighty warships—"

Kalen belted out a disturbing laugh.

"A single Goro'Sai is more than you'll ever need," he said, leaning back in his chair. "The *Shadow People* have awakened, Mr. Winters," he said, causing the temperature of the room to drop further. "Do not force us to leave Andromeda to address this matter. For none in your galaxy will endure our wrath."

Before Teric could speak, everything in the room went black.

#

Inside his grand hall, Teric remained pinned to his chair, eyes wide open and mouth agape. His limbs twitched as if besieged by powerful currents of electricity. With labored breathing, and a brow drenched with sweat, Teric had the look of a man moments away from cardiac arrest. Samson continued to watch, wondering if his old friend had finally gone too far. They'd been down this road before. He told Teric that allying with such vicious beings was a mistake, but Teric was convinced there was no other way to see the destiny of mankind fulfilled.

Samson had faithfully served at Teric's side for hundreds of years since the start of the Great Purge. He remembered it well, the day he and the ISF Armada under his command defected to Orion's Shield, giving Teric's beloved movement a much-needed boost in military might and training. For his tremendous sacrifice, Teric bestowed upon Samson the same life-extending bio-enhancements from which he personally benefited, so that Samson might command his forces for centuries to come.

However, those enhancements came at great price. Samson's body partially rejected the bio-treatments, causing extensive nerve damage in his left leg. The injury was so severe it prevented subsequent cybernetic enhancements from fully integrating with his body, leaving him with the distinctive limp for which he was so well known.

As Samson thought upon all his body had endured for that great cause, his leg began to ache, prompting him to get up and stretch. As he moved, he could hear the whirring and whining of the servos in his partially cybernetic leg. But as soon as he had risen from his seat, Teric snapped out of his immobilized state.

"They've got to come up with a better way of communicating," said Teric, struggling to shake off the effects of the mental transference.

Samson was pleased that Teric still had his wits about him. But still, he didn't like this arrangement.

"Ever since they implanted that... thing in your head, there's been nothing but trouble," said Samson, genuinely concerned. "One day it's going to kill you."

"A small price to pay," replied Teric. "We're so close." His eyes widened with a rare look of excitement. "They're finally intervening on our behalf."

Samson was shocked, having always felt the arrangement was a little too *one-sided* for his tastes.

"How so?" asked Samson, skeptical.

"He said they're sending something called a Goro-Sai," said Teric, trying to get the pronunciation just right. "They've never mentioned that word before. But at this point, we'll take whatever they're willing to give."

"I'm telling you, Teric. We need to cut ties with these people immediately," pleaded Samson. "I left the ISL all those centuries ago because I believed in your father's message," he continued, deciding it was time to finally speak his mind. "Your father's vision was to see a universe where mankind could thrive without Xeno interference. And now, you secretly align us with an unknown alien race. How can you justify—"

"My father..." said Teric, the very words bringing a sour taste to his mouth. "*Zedolph the Visionary* they called him," he said, irritated that after all these years he was yet unable to escape the shadow of his father. "Well, let me tell you something about *my father*."

But before Teric could continue, the temperature in the room suddenly dropped, as if a freezing cold wind suddenly penetrated the metallic walls of his grand hall.

In unison, Teric and Samson turned their heads to see a dark, shadowy mist forming in a corner of the room. The high-pitched shrill of the mysterious voices returned, only this time Samson could hear it as well. And from the dark mist, *she* appeared. Standing over six feet tall with a slender yet muscular build, she was adorned in a formfitting black armor so well made it never hindered her graceful stride.

From across the room the two men studied her every movement. As she walked, they could see an advanced, pistol-like side arm strapped to her left leg. On the other was something like a spiked whip, though seemingly much shorter than traditional whips still in existence.

The fullness of the figure's armor and weapons was obscured by a black, hooded cloak flowing behind her as she quickly moved across the pristine floor. As she moved closer, they could make out the intricately engraved hilt of what was surely a sword, about the shape and size of an ancient Japanese katana, fitted with a beautifully curved hand guard. She stopped several feet short of Teric and Samson, casting an ominous glare at the two men. But as close as she stood to them, her face seemed shrouded in a dark mist, allowing them to see only the purplish-black glow of her cold, pupil-less eyes.

After a bitter silence, Teric spoke.

"Greetings, my dear," said Teric, not knowing how to address the menacing figure standing before them. "And what, may I ask, do they call you?"

"My true name," said the woman, "cannot be uttered in your primitive tongue. But you may address me as *Xeela*."

As she spoke, Xeela pulled back the hood of her dark cloak. As she did so, the black mist shrouding her face slowly dissipated. She was hauntingly beautiful. The right side of her slender face was partially covered by shoulder-length jet-black hair, braided and bound with spiked metal rings.

The hair on the left side of her head was shaved to near scalp level, revealing a narrow cybernetic plate running from her temple to the back of her head. Her face was pale, covered in small scales, only noticeable when standing up close. Stretching across her face like an eye mask was a shadowy mark, like a tribal tattoo, causing the fiery purplish glow of her eyes to stand out even more.

"I am, Goro-Sai," said Xeela, casting her unnerving glare at the two feeble humans standing in her presence.

"Lord Kalen mentioned that word, Goro-Sai," said Teric. "What does it mean?"

Though obviously irritated, Xeela indulged him. "The closest, albeit limited, translation you have would be... Death."

Teric was intrigued with everything about Xeela, especially her claim to be the very personification of death itself. But as he turned to look at his old friend, he could see that Samson, though somewhat disturbed by her presence, didn't share the same enthusiasm.

Being a former high Admiral, Samson believed only in the unbridled firepower of a well-armed fleet coupled with the destructive capabilities of a highly trained ground force. He never believed that one person alone could make a difference. To him, strength was found only in numbers... and nothing else. He made an about-face turn and stormed toward the door, refusing to acknowledge Xeela's presence.

"Samson," called Teric, "where are you going?"

Samson stopped in his tracks and turned toward them.

"I'll have no part of this... foolishness," yelled Samson in an uncharacteristic outburst of emotion.

As Xeela stood watching with an expressionless look on her face, Teric moved toward Samson. As much as Teric had a good feeling about Xeela and her capabilities, he had no desire to move forward without the trusted counsel of his longtime friend.

"We need you, Samson," said Teric.

"No, we do not," said Xeela, sharply. "Let your spineless dog scurry back to his cave. We have work to do."

Furious, Samson turned to address Xeela directly from across the room.

"What did you say, alien?" barked Samson. "You're an even bigger fool than I thought. You must be crazy if you think you can stride in here with your primitive weapons and fancy armor..."

As he spoke, Xeela removed the spiked whip attached to her leg and lashed it toward Samson with blinding speed. As the whip extended, its length increased beyond its original limits, allowing it to easily reach its target. But just before wrapping itself tightly around Samson's neck, the spikes adorning the whip retracted, keeping Xeela from completely severing his head in the process.

Samson groaned as the whip slowly tightened itself around his neck like a powerful python. It was clear that the whip was a living entity, capable of its own movement, yet completely obedient to Xeela's telepathic commands. She yanked the whip with an unexpected force that sent Samson flying several feet toward her, sending him careening to the ground. With a flick of her wrist, a sudden and powerful electrical current radiated through the whip, causing Samson to convulse violently until blood streamed from his ears. With another flick of her wrist the electrical field dissipated, after which the whip began to retract itself into the hilt, dragging Samson across the floor.

He looked toward Teric to intervene. But Teric simply watched in amazement... but did nothing. As Samson lay at her feet, Xeela released him from the crushing confines of the whip and returned the weapon to the attachment on her leg.

Samson staggered to his knees, coughing and wheezing, struggling to breathe. Moments later, Xeela grabbed him by his collar, yanking him off the floor as if he weighed little more than a child. Her purple eyes burned bright like an unnatural fire. As she stared at him face to face, the dark crystals imbedded in her armor glowed in the same manner as her eyes, casting its surreal purplish-black radiance into the room.

"The next insolent word to part your lips... will be your last," said Xeela.

Samson conceded, nodding his head in acknowledgment.

Seeing that her point had been made, Xeela dropped the former high admiral to the ground and moved toward Teric's large desk at the center of the room.

As she did so, Teric continued to watch her every movement. He snapped out of it, rushing to Samson's side to help. But Samson angrily waved off Teric's assistance, deciding to stand on his own. He struggled, but eventually made it back to his feet.

Teric tried to offer his apologies for not coming to his friend's aid, but Samson returned a hardened look that he had never seen in all the centuries they've known each other. In sorrow, Teric watched as Samson limped toward the table where Xeela awaited so they could plan their next move. But he quickly shook off the feeling, resolving within himself that fulfilling his destiny was all that truly mattered.

"Come, Mr. Winters," said Xeela from across the room. "We have much to discuss."

Teric took a deep breath and exhaled slowly. As he approached his desk, Samson glared at him in a way that he had never seen before. And in that moment Teric knew that on many levels, things would never be the same. After joining Xeela at the desk, Teric paused, looking at his two companions for a moment. He then opened his mouth, deciding to deliver a longwinded speech about how it was *destiny* that brought them together.

But little did Teric know, Xeela had other reasons for revealing herself, for her true mission was twofold. And as she listened to Teric pontificate about his vision for humanity, she thought upon her people's greatest prophecy, which was soon to be fulfilled... a prophecy that would impact every life form in the Milky Way forever. As she stood silently, a singular thought echoed throughout her mind, *the time of the eighth and final Void Walker... is at hand.*

CHAPTER FORTY-FIVE

The headquarters of the Interstellar Guard's Special Warfare Division was a massive structure located on the southeast corner of the Citadel grounds in the heart of Earth's Haven City. From its exterior, the circular-shaped, multi-story building looked unremarkable. But inside the walls of that highly secretive facility resided the entire command chain for the ISG's most elite special forces detachment, Saber Brigade.

At the rear of Saber HQ, Donald Shepard's third-story office window overlooked the vast training grounds below. From his vantage point, he could see scores of soldiers running live fire exercises in the distance, including simulated Starbase seizures and marauder drop-ship raids on heavily fortified structures. From his position, Donald could hear the familiar sounds of newly inducted Saber team operators singing cadence as their drill sergeants took them on the twenty-five-mile run around the perimeter of the main training grounds. Failing to complete the grueling run meant expulsion from the Saber program altogether.

It seemed like only yesterday when he was a young recruit, hungry to prove himself worthy of bearing the coveted Saber Team patch displayed proudly on his left shoulder. The patch itself was a medieval heater shield, upon which was an image of a skull resting atop two crossed, downward-facing sabers. Beneath the blades was the Saber Team Unit number followed by the brigade motto written in Latin across the bottom of the shield, *Fatiscat per Scutum.*

The motto meant *Splitting the Shield,* a fitting phrase, as the Saber Brigade was formed to be the edge of the blade that would rend the Orion's Shield terrorist organization to its core. But the last mission was especially tough on Donald, having lost so many fine soldiers under his command at one time. He played the event in his mind over and over, coming up with at least half a dozen things he could have done differently. But such was the life of a commander; once you make the call, you must live with the consequences, no exceptions. As he thought back on the days before he was responsible for so many lives, Donald closed his eyes and sighed, reminiscing on the days when he was simply taking orders instead of giving them.

As Donald continued to watch the troops train below, Lieutenant Alex Chavez entered his open office door. He held a beautifully engraved wooden case that housed medals for the soldiers slain during the battle on Protos IV. Of course, the families of the soldiers were fed a fictitious story surrounding the death of their loved ones, after which the soldiers would be laid to rest following a grand military funeral worthy of the distinguished service they rendered to the Interstellar League. At the end of the elaborate ceremony, the families would accept the posthumous medals of valor on behalf of the dearly departed.

"Sir," said Alex as he placed the box of medals on Donald's meticulously organized desk, "the funeral arrangements are set. Major Ellis wants to know if you're ready for her to notify the families."

"No," said Donald following a long pause. "I'll do it myself."

Alex wasn't surprised, knowing the type of man Captain Shepard was. He wasn't like the rest of the commanders, willing to send others to carry out the less glamorous aspects of the job. It was those attributes, and more that Donald possessed, that earned Alex's unfettering respect and loyalty.

"I'll let her know immediately," said Alex.

"You have the intel I asked for?" said Donald, ready to get down to business.

"Yes, sir," said Alex sliding a black holo-file across his commander's desk.

Donald returned to his seat and activated the file with a quick tap of the device. A moment later, a full dossier of Marcus La'Dek projected in front of them. After skimming through the first page of the file, he looked up at Alex.

"So, what's our boy been up to?" asked Donald.

"Not much actually," said Alex. "He's been out for over a year now, but he's kept it clean the entire time. His name hasn't showed up on a single incident report in ISL space or abroad."

"I told them it was a mistake lettin' this guy out," said Donald under his breath.

"Excuse me, sir?" said Alex.

"Nothing," said Donald. "So, what would make a guy go from zero to a hundred like this?"

"Not sure, but I did a little research," said Alex taking the empty seat in front of the desk. "He's been working for his family's cargo transport business on the Stratus one Star Port. Apparently, his sister took it over after their father died about four years ago."

"What do we know about the business?" asked Donald while swiping through the holographic pages of the dossier.

"Just a run-of-the-mill discount shipping company. But they're in serious financial straits from what I've been able to gather."

"Money troubles, you say," said Donald, finally deactivating the holo-file. "That's how it always starts."

"My thoughts exactly, sir," said Alex with a slight smile breaking the corner of his mouth. "What's our move?"

"Perhaps it's time we pay Ms. La'Dek a little visit," said Donald, rising from his chair.

"Sounds good," said Alex, standing to his feet as well. "I'll prep the ship."

CHAPTER FORTY-SIX

"This... is a lot to take in, Marcus," said Samantha, her face filling the view screen of *The Indicator*'s main communication's panel.

"I know," said Marcus, barely able to look his sister in the eyes. "I should've been straight with you from the beginning."

It had been twenty-four hours since he dumped Daren and the crew on the Aris Star Port. He was in no hurry to go home and face his family. So, he took a lengthy detour to a nearby binary star system that offered one of the most majestic views in the known galaxy, just to clear his head. After ignoring several calls from Samantha, he finally decided to reach out. And for the last hour, he bared his soul to her. As painful as it was, he felt a sense of liberation as he unloaded every weight that had been holding him down.

Since reuniting with her brother back at Runner's End, Samantha had always given Marcus a wide berth when it came to the details of his past. Part of her wanted to know everything, but an even larger part thought it best to remain in the dark. As she sat in her private office inside La'Dek Transports, Samantha cringed as she learned the gritty details of her brother's *other* life. The entire conversation swept her into a wild emotional vortex leaving her breathless and heartbroken.

Marcus told her everything. He told her how he and Daren first met, and of the brutality they experienced under the oppressive hand of Daren's father, Jax West. He told her of the crew that had become his second family in the years he'd been gone. He even told her how he truly felt about the one and only love of his life, Skyela Evans. He then moved into recent events, confessing the truth of what Daren wanted when he first walked into La'Dek Transports.

The sheer weight of the conversation was almost too much for Samantha to handle. Part of her wanted to reach out and hug her brother, while another part of her wanted to slap him for being so careless. More than that, she wanted to kick herself for not listening to her gut feeling about the job in the first place.

"It's my fault too," said Samantha amid a sigh of frustration. "I knew somethin' wasn't right about this cargo run from the beginning. But I signed off on it anyway," she admitted. "I think we were both desperate for the money."

"I guess so. But I'm tellin' you, I'm done with the old life," said Marcus. "For real this time."

"I believe you," said Samantha with a radiant smile on her face. "But it must have been hard, dumping your old crew on Aris like that," said Samantha. "And as hard as it is to accept, they are part of you. They're as much your family as we are. You sure that was the right call?"

"What else was I supposed to do?" said Marcus, irritated as he thought back on everything that had transpired over the last few days. "I should've left them in the past, all of 'em. And now, it's only a matter of time before that decision blows back on me *and* our family."

"I'm not saying you should have continued the job or anything like that," said Samantha, attempting to clarify. "Maybe you could've talked them out of—"

"They ain't my problem anymore," said Marcus. "The kind of people they're involved with, won't let you just walk away. Like I said, I'm done with this—"

Before he could finish, an emergency alert blared throughout the bridge of the ship. The call came through the secondary comm system. Marcus answered the call without disconnecting the transmission with Samantha, leaving her to hear the entire conversation. He could barely hear the caller over the sounds of explosions and weapons fire sounding off in the background.

"It... was a trap," yelled the voice on the transmission. "They took him!"

"Skye?" yelled Marcus, finally recognizing the distorted voice on the other end. "They took who? What's goin' on?"

"They took Daren," yelled Skye as loud as she could into her comm unit. "Lock onto our position," said Skye. The deafening sound of her pulse rifle going off forced Marcus to cut the volume down. "Come get us, please!"

Marcus hesitated for a moment, then responded. "You got yourselves into this, now figure it out!"

Before she could reply, he heard her gut-wrenching scream just before the transmission died. He sat staring at the secondary comm unit for a moment, wondering if the hurtful words he just spoke to Skye would be the last she would ever hear. While suddenly overcome with dread, Marcus did his best to suppress the feelings, knowing that he had to take care of his real family now.

He turned to Samantha who was still on the main view screen. In that moment, the reality of Marcus' world suddenly became very real to her. Being a former soldier, Samantha was appalled at the words that came out of her brother's mouth.

"I know all those years of hard livin' ain't made you that cold," said Samantha, knowing that the true heart of her brother hadn't totally turned to stone.

"Like I said, things are different now," said Marcus, trying to convince himself that abandoning them was the right thing to do.

"You need to stop it, Marcus," said Samantha, not even letting her brother continue. "You know you still love that girl," she said, cutting straight to the chase. "And I don't care what you say, you still care about your crew. Even, Daren."

"Look... all I'm worried about is Liz, Kayla, and you, that's it. You guys are all the family I have left," said Marcus. "Daren ain't even—"

"Blood?" said Samantha, completing his sentence. "Does it really matter that the same blood doesn't run through your veins?" she continued. "Like it or not, *you're* brothers. And yes... you are your brother's keeper."

Marcus hated that she was right. He couldn't shake the inescapable truth that he *did* care. He cared about all of them... even Daren. But he was tired of the vicious cycle. Daren messes up, and Marcus is right there to clean it up.

"How many times do you have to get burned before you cut 'em off, Sam?" said Marcus. "I can't keep doin' this."

"Good or bad, you can't choose family," said Samantha. "Sometimes... you just have to roll with the punches... That's what we do."

"But y'all have to get out of there. It's only a matter of time before Donald shows up—"

"We can't just pack up and leave. You know that," said his sister. "Let me worry about Donald. Right now, the rest of your family needs you," she said. "Skye, needs you."

Marcus knew from the start what he had to do, which is why he had quietly started the trace as soon as he realized it was Skye. At the end of the day, there was no way he could ever turn his back on her, knowing her life was in danger. Perhaps he just needed a moment to vent... and to hear his sister say it was okay.

As they sat in silence, he continued tracing the signal back to its source. Moments later, the coordinates were locked. He looked up at his sister.

"It's alright, Marcus. We'll be fine," said Samantha, touching her hand to the view screen. "Stay safe out there, baby brother. And do what you do."

Marcus gave her a smile, then initiated the ship's hyper-jump engines. Within seconds, *The Indicator* made the jump to hyperspace.

#

As *The Indicator* entered hyperspace, the transmission died, leaving Samantha staring at an empty view screen. She knew that Marcus would never forgive himself if he walked out on his old crew, and Samantha could never forgive herself if she let him. Like she said, *You can't choose family.* The crew was part of him, and she had to accept that. For the last ten minutes, she had been ignoring the incoming message coming from Kayla. She finally answered the call.

"Yes, Kayla. I'm here," said Samantha over the comm system.

"I didn't mean to blow up your comm unit like that, but you have two visitors here. And they look serious," said Kayla.

Samantha knew exactly who the two *visitors* were.

"Go ahead and send them in," Samantha replied to her cousin.

She terminated the communication and took a deep breath.

Just roll with it, Sam, she told herself.

Moments later, the automated doors to her office opened. She looked up to see two stern-looking ISG officers, one a Captain, the other a Lieutenant, both decked out in full military dress uniforms, complete with a chest full of ribbons and metals shined so bright, she could almost see her reflection in them. The sight of them brought back memories of her time in the military.

"I'm, Captain Donald Shepard," said the officer, extending his hand, which she greeted in kind. "This is my second in command, Lieutenant Alex Chavez."

Samantha shook Alex's hand as well, after which she gestured for them to have a seat in the two empty chairs facing her desk. As they all sat, Samantha said, "Captain Shepard, Lieutenant Chavez. Welcome to La'Dek Transports. How can I help you? You know, we've been trying to land a military contract for months."

"I'm afraid we're not here about a shipping contract," said Donald with a slight smile.

"I see," said Samantha, knowing full well what this visit was about. "Then what *can* I do for you?"

"We need to discuss your brother, Marcus La'Dek," said Donald, deciding to get straight to the point.

And there it was. She'd dealt with men like Donald and Alex during her time in the Interstellar Guard. She knew full well they wouldn't leave her office until they got the answers they wanted.

"My brother?" said Samantha, trying to act surprised. "Why? Is he in some sort of trouble?"

CHAPTER FORTY-SEVEN

After Daren and his crew arrived on Xenon Prime, everything went sideways. Between Tommy's sketchy research, his unreliable tracker, and a tip they had received from one of his shady information dealers, they found themselves on the fifth planet of the Xenon System, the planet Tommy was sure Goran and Sarrus would hit in search of the only person alive with the scything skills to crack an Mk17, the programmer-turned-fugitive, Bradley Jennings. But within hours of their arrival, the team soon learned that the carefully planted info trail they'd been following the whole time was the brainchild of none other than Goran himself.

Goran knew that Tommy was, at his core, a conniving opportunist, out for only himself. He wagered that once they ran off with the Benton Chamber, Tommy would likely go crawling back to Daren to somehow convince him to join forces, hoping to retrieve his coveted prize, which would lead Daren straight to Xenon. And even if Tommy didn't reunite with his former business partner, they would easily manipulate him into luring Daren into their well-placed trap. Either way, they would have the only man with the access codes to the Mk17, knowing that Bradley Jennings had long since moved on from the Xenon system. In the end, Daren was always their only play. And fortunately for Goran and his brother, Sarrus, Tommy made their life much easier by acting in such a predictable manner.

And soon after Daren and his crew's arrival on the planet's surface, Goran ordered their forces to strike with explicit orders to kill all except for their sole target, Daren West.

The first wave of the attack had been brutal, resulting in Daren's abduction during the initial assault. The team barely escaped with their lives, though they managed to decimate many of their attackers, sending them limping back to their bosses, where they would likely be dealt with harshly.

"We need to fall back and regroup," said Tony reluctantly following the exhausting skirmish. "I saw an abandoned house earlier. Should provide some solid cover in case they counterattack."

"You're right," said Skye, frustrated that they were in no shape to press forward. "Let's move out," she said to the team.

When they reached the house, Tony rushed in to clear the incredibly sturdy structure then took up a firing position at a window on the south side. Moments later, Tommy dragged Jason to a tattered couch in the corner of what used to be a living room. He was in bad shape and unconscious. Skye followed closely behind, doing everything she could to stabilize her brother. The entire situation brought back disturbing memories of the day they almost lost their beloved team member, Maxlyn Wesner.

Thankfully, Jason was nowhere near in as bad a shape as Max. All his limbs were intact, though he was severely bruised and scarred following a rocket attack on their position, the same blast that cut short her communication with Marcus. She never did catch the words that Marcus said over the sound of the explosion, but she prayed that he somehow got the message and was en route to pick them up.

"We're runnin' low in the ammo department, guys," yelled Tommy to the crew shortly after taking a position near the door on the north side of the house. "At this rate, we're gonna be throwin' rocks if we don't get to our ship and rearm."

But they all knew it was wishful thinking. By now the ship they had rented on Aris was surely a smoldering scrap heap. There was no way they were leaving that planet alive, short of some divine intervention. They just weren't prepared for a sustained firefight. At best, they expected a shootout with Goran, Sarrus, and perhaps a few others. They had no idea they'd run into what turned out to be a small army. Not quite the size of Tommy's forces back on Protos IV, but it was large enough.

"Not sure we can hold off another wave," said Tony, reloading his weapon. He noticed Skye in the corner trying desperately to reach Marcus on her comm unit. "Why do you keep foolin' with that thing? He ain't comin'. It's just us now."

"He might," said Skye, trying to hold out hope. She decided to give the comm unit a rest, returning it to one of the pouches on her belt.

"I'm headin' to scout the area," said Tony. "Maybe I can get an idea of where they're holdin' Daren."

"Sounds good," said Skye. She looked at Tony with a wary eye. "But the first sign of trouble, you fall back. Don't engage by yourself, you hear?"

Tony sighed in frustration but nodded his head. He longed for a challenge such as the one they were facing, and it took everything within him to suppress the urge to seek out the enemy on his own. It was a genetic trait of the Gorean Cyborg. They were bred for battle, and longed to enter their terrifying natural state, known to outsiders as *The Blood Rage.* For the Gorean, life was a delicate balancing act as they struggled to suppress their overwhelming killer instincts. To aid in this, Goreans relied on a synthetic liquid called *granacite,* used to help keep their emotions in check.

Gorean Cyborgs could lower their granacite levels at will, giving them an extra boost in strength, reflexes, and speed. But dropping the levels too low would send them into the blood rage, turning them into a near unstoppable force, while at the same time making them a danger to both themselves and those around them. Tony knew that his duty to the team came first, so he had no intention of going against Skye's wishes. He grabbed his gear, stabilized his granacite levels, and headed out the door.

Skye looked over to Tommy, who was staring out the doorway toward the open field in front of them, beyond which lay a city in ruins. He was looking for any sign of movement. Once a vibrant manufacturing world within the Atrix sector of the Outer Core, Xenon Prime was now little more than a shadow of its former glory, stripped of much of its natural resources; the result of generations of over-mining by mega rich corporations and their countless subsidiaries. With scores of abandoned mega cities and smaller towns littered throughout every landmass, Xenon Prime was the perfect hiding spot for those wishing to lie low, not to mention the ideal location for an ambush.

He had never been so careless. Tommy was furious with himself for falling for the simple *distract-and-grab* move employed by Goran and Sarrus, the elementary tactic that caused the Benton Chamber to slip right through his fingers. And by failing to vet the information coming to him from his contacts, he had just marched everyone headlong into what was obviously a trap.

He let out a long, slow sigh as he continued to stare across the field. With the aid of his enhanced ocular implants and tactical shades, he observed what used to be a sprawling metropolis called Solis City. But Solis, like many of the structures still standing on the planet, was little more than a ghost town that nature was well on its way to reclaiming. Every crumbling building, including the house in which they were hiding, was overgrown with vegetation, some of which had become natural habitats for the local wildlife.

It was a dreary scene; the absolute last place that Tommy expected to die. And just when the depressing landscape couldn't get any worse, it began to rain, hard. Tommy shook his head in disbelief as he watched the powerful thunderclouds swelling above. The sensors in his tactical shades confirmed that conditions were expected to worsen over the next few hours. It was about to be a *very* long day. As Tommy continued scanning the horizon, Skye approached from his left. He heard her footsteps, but never turned his gaze from the open landscape outside their little hideout. As she stood next to him in silence, Tommy cleared his throat to speak.

"Looks like a storm is comin'."

Skye looked out the door to see the harsh weather conditions forming. But she knew the rain would be the least of their troubles.

"Yeah, tell me about it," she replied.

"Hopefully it'll slow 'em down."

"Hopefully."

"The way I figure, if we could just—"

"Thank you, Tommy," said Skye. It took a lot for her to utter those small words. But she meant what she said.

For a rare moment, Tommy was at a total loss for words.

"You know, for what you did for Jason back there," she continued. "He'd be dead if it weren't for you."

Tommy thought nothing of his earlier actions. Following the rocket blast that knocked Jason cold, he instinctively fought back the thugs attempting to finish her brother off. At the end of the day, Jason was part of the team, and Tommy figured they needed every shooter they had operational. But still, she caught him off guard by the rare display of gratitude, which he never expected from someone as emotionally hardened as Skye.

"Don't worry about it. I'm sure your brother would've done the same for me," he said, attempting to lighten the mood a bit.

After a moment, the two started laughing.

"Yeah, I'm... pretty sure he would've left you to bleed out in the gutter without a second thought," said Skye honestly. Then, with a straight face, she gently removed his shades and looked directly into his pale eyes. "But I thank you anyway," she said, motioning toward her unconscious brother in the other room, "on *his* behalf."

She returned his tac-shades and moved to check in on her brother. After confirming he was okay, she took up Tony's position on the south side of the house. As she walked away, Tommy watched her for a moment. He then donned his tactical eyewear and turned his attention back to the dreary Xenon landscape.

Within twenty minutes, Skye noticed Tony in the distance, sprinting toward the house. It wasn't a good sign. As Tony stormed into the room, he stopped momentarily to catch his breath. Being true cyborgs, Goreans were not total robots. They were largely organic. They felt emotion, pain, and fatigue; they just had greater thresholds than the average humanoid. After quickly gathering himself, Tony turned to the team.

"They're comin'," said Tony, prepping his gear for battle. "They're spreadin' out. Expect them to hit us from all sides."

"How much time do we have?" asked Tommy.

"Maybe ten minutes," replied the cyborg.

"So, this is it, fellas," said Skye as she too began prepping her weapons for the coming storm.

Everyone nodded in agreement, deciding they weren't about to go down without a fight. As they readied their gear, Jason staggered toward the team from the living room.

"We doin' this or what?" said Jason, jamming an energy cartridge into the rear of his rifle. He had no intention of sitting this one out.

Immediately, the combat medic in Skye kicked in.

"You're in no condition to—"

"Get to your position," said Jason to his sister. "It's 'bout to get ugly."

One look at her younger brother, and Skye knew that he was going to fight, no matter what. He was just like her in that regard, so she backed off. She reached into her med-pouch and tossed her brother a pain injector. She then took up a firing position near the south window.

Jason injected the meds into the side of his neck. While it didn't stop the pain altogether, it did take enough of the edge off to allow him to fight.

They all moved to strategic locations throughout the house, covering all approaches. Minutes after settling into their firing positions, the enemy crept in over every horizon. To their dismay, the team was surrounded by what look like a Dragen Alliance horde, all armed with melee weapons and short-range blast pistols, all eager to stain the soil of Xenon Prime with the team's blood. And without warning, they attacked, rushing toward the house in a blind rage.

With no fire order needed, Skye and the rest of the team opened up on the oncoming horde, thus beginning what would surely be their final stand.

CHAPTER FORTY-EIGHT

Aided by the beat-up maintenance droid he kept aboard *The Indicator* for hauling heavy loads, Marcus spent the hour-long trip to the Xenon system working his fingers to the bone, trying to get *The Indicator*'s weapon systems online. Though partially dismantled so they wouldn't register on star port security scans, Marcus still had a few weapons aboard *The Indicator* in case of an emergency. He had two pulsar rapid-fire cannons, four rocket pods, and a complement of self-guided hornet missiles. The ammo was stashed away in hidden compartments, wholly invisible to external security sensors.

The armament he carried was part of a backup covert defense system Tony had worked on years ago, but never actually finished. After reclaiming his ship from Jason more than a year earlier, Marcus intended to complete the install, but never could successfully bring the decades-old system online, nor did he have the time to sit down and troubleshoot the myriad of problems identified on routine diagnostics scans. So, the formidable weapon system remained within the belly of *The Indicator*, collecting dust. But with what he was likely to face on Xenon Prime, Marcus had no choice but to get those weapons operational in record time.

While the maintenance droid reassembled the missile pods and cannons, Marcus pored over terabytes of technical manuals and research data until he finally figured out the problem. It took every minute of the trip, but he managed to get the weapons systems working, though he would have no time to conduct a thorough test. He wasn't even sure if the batch of missiles he had on board were all in a serviceable state, but he would find out soon enough. As he finished running the final diagnostics on the defense system, *The Indicator* arrived in the Xenon system in a burst of multicolored light.

He pushed the ship harder than ever before. For a star system so remote, he would normally have plotted a course using a series of sequential hyper-jumps, giving the drives ample time to cool and recharge before reactivation. But that was a luxury he simply couldn't afford. He made the trip to Xenon using a single *omega-jump*, a dangerous technique, ill advised by just about every starship engineer across the galaxy. He was certain the damage to the drives would be extensive, but he was confident *The Indicator* could handle the stress. And as expected, the ship held together, but barely.

When he arrived at the outskirts of the Xenon system, Marcus initiated an emergency shutdown of the hyper-jump engines, hoping to slow the rapid overheating and excessive energy buildup occurring in the engine room. Using every ounce of electronic trickery gleaned from Jason and Tony over the years, Marcus managed to stabilize the rapidly declining conditions aboard the ship. After rerouting climate controls and several other subsystems, he managed to get the internal drive temperatures to decrease at a steady rate. He also managed to stabilize ship-wide energy levels, which were moments away from going critical.

He breathed a sigh of relief, knowing he just prevented both the engines and the entire ship from exploding in space, which would have brought about an unfortunate end to his short-lived rescue operation. After confirming the stability of *The Indicator*'s propulsion system, Marcus maneuvered the ship into an equatorial orbit of Xenon Prime. He then patched into what was left of the planet's failing satellite network. He didn't have much to work with, but he was able to triangulate the team's position. It wouldn't be one hundred percent accurate, but it was better than nothing, as he was no longer picking up a signal from Skye's communicator.

With the coordinates locked, Marcus engaged maneuvering thrusters, sending *The Indicator* thundering toward the planet's upper atmosphere like a meteor. Once planet side, he realized that he was way off course. He used the ship's mid-range sensors to get a more accurate readout of the team's location and of the situation on the ground. It wasn't good. He was picking up multiple targets converging on a small building that was no doubt housing the team. He adjusted course and set the planetary thrusters to maximum.

In seconds, *The Indicator* rocketed toward the team's location. As the ship rose to incredible speeds, his anxiety levels seemed to rise just as fast. He couldn't get Skye's soul-crushing scream out of his head. He pushed the ship beyond its limits, hoping to make it in time.

#

The situation on the ground was a complete mess. The only reason Skye and her team were still alive was because Sarrus was forced to send a wave of poorly equipped forces to rush the compound, armed with only short-range firearms and hand-to-hand weapons. But the initial assault forced Skye and the others to burn through the bulk of their remaining ammo, so it wasn't a total loss. In the end, the first wave was little more than cannon fodder. Skye and the crew took them down with little problem. But the second wave wouldn't go down so easily.

Though fitted with personal shield generators, the extra layer of protection wouldn't make the second wave of troops totally invulnerable, but Sarrus figured it would allow at least some of them to get close enough to storm the building. But still, Sarrus was surprised that he even had to resort to a second wave. And after that last failed assault, one thing became abundantly clear, while the walls of that building remained intact, they would have no chance at taking out their targets.

"Time to soften up their little nest," said Sarrus to one of his fire team leaders, a human calling himself, Gromm. "Is the RAT operational?"

"My guys are still working on it," said Gromm.

"Get that tank up here now," growled Sarrus to his fire team leader.

Gromm acknowledged and rushed to the rear to help his team prepare the barely functional Robotic Autonomous Tank for battle.

Fortunately for Sarrus and Goran, there was already a small Dragen Alliance cell, comprised mostly of Lyrians and a few humans operating out of Xenon Prime. And though many of their forces were engaged in off-planet operations, along with every space capable ship owned by the cell, there were enough of them left to assist in the operation. And while they weren't the most refined warriors that Sarrus had ever commanded, they all relished the opportunity to prove themselves to Ar'Gallious, willing to lay down their lives for the cause if necessary.

And they would indeed lay down their lives, as Sarrus made it abundantly clear that Ar'Gallious would accept nothing less than perfection. And given the Dragen Alliance leader's reputation, they all knew it was better to die on the battlefield than to be brought before Ar'Gallious for failure. However, Sarrus' biggest challenge was overcoming just how ill-equipped the forces were. All the cell's best soldiers and equipment were off planet, leaving Sarrus to command what amounted to the "B-Team," armed with substandard weapons and equipment.

Besides the poor quality of troops, Sarrus was still seething over the loss of his ship, the vessel that was irreparably damaged by the Gorean Cyborg on Daren's team during their initial ambush, grounding it permanently. And with no space-capable ships available planet side, they were stranded until reinforcements would arrive, which was still more than an hour away. Therefore, Sarrus had one job, to keep the rest of Daren's crew from mounting a counteroffensive, at least until their transport arrived.

As Sarrus continued to glare at their target in the distance, the twelve-foot bipedal tank finally made its way to the front lines. The ground shook as the metallic beast negotiated the muddy Xenon soil. While the RAT they fielded was an older, poorly maintained model, it was fitted with an upgraded artificial intelligence neural network, allowing Sarrus to give commands to the tank as if he were speaking to a living being.

"Open fire on that building," Sarrus yelled to the RAT. "No one survives!"

"ORDER ACKNOWLEDGED," belted the RAT. The walking tank moved forward several meters then deployed the two energy-based Gatling guns on the upper left and right sides of its chassis.

Tommy moved to the north window of the building to get a better look. "They got a RAT!" he yelled to the crew. "Everybody down!"

Seconds later, the RAT unleashed a barrage of energy bolts into the north wall, quickly reducing it to rubble. Inside the building, the team low-crawled away from the northern section. As they moved, they could feel the heat of the rounds streaking above their heads. As the wall crumbled, a portion of the roof collapsed.

It was a remarkable spectacle. The Dragen Alliance mercenaries laughed as the RAT made short work of the sturdy walls protecting the enemy. But shortly after unleashing the hailstorm of rounds into the building, the RAT's guns started to overheat and misfire.

"Cease fire," yelled Sarrus, exasperated by the useless piece of junk.

The RAT immediately disengaged, belting out a series of warning messages regarding the status of its primary weapon.

Sarrus examined the building with his portable bio-scanner. To his dismay, everyone survived. He was hoping to bring the entire building down on their heads, but it didn't happen. And with one look at Gromm as he feverishly worked on the RAT to get it operational, it was clear that his best chance to end the ordeal early was a complete failure. He had no choice but to send in the second wave, hoping they could at least storm the building and engage the enemy in close-quarters combat. "Fire Team Two, move in," yelled Sarrus to the last of his forces.

#

Inside the building, Skye and her team managed to dig themselves out of the rubble, only to find their position moments away from being overrun.

"We got incoming," yelled Skye. "Take 'em down!"

As the second wave rushed the building, they were met with relentless fire.

Noticing their energy rounds were being absorbed into their targets, Jason yelled to the crew. "They have shields!"

Tony set his heavy repeater to maximum charge, knowing that his weapon would only get off a few rounds before completely fizzling out. With a smooth sweeping motion, Tony mowed down the initial wave of their attackers, overloading their shield generators, literally cutting them in half. While able to squeeze off a few more rounds than expected, Tony's weapon, like that of the rest of his team, soon ran dry.

As the surviving Dragen Alliance mercenaries stormed the building, Tony flipped his heavy repeater into the air, and grabbed it by the barrel. He swung the weapon like a club, knocking many of his attackers off their feet. The rest of the team dropped their rifles and met the oncoming attackers in hand-to-hand combat.

#

Marcus thundered onto the scene shortly after the Dragen Alliance thugs rushed the building. Those who had not yet made it inside ran in fear at the first sight of *The Indicator*. He immediately locked onto the RAT, firing a barrage of hornet missiles at the metal behemoth. But some of the untested rockets malfunctioned, failing to launch from their tubes. Still, he was able to release enough to get the job done.

Sarrus looked up only to see the twisting smoke trails of what had to be twenty or more projectiles heading his way. He immediately dove for cover.

The RAT picked up the incoming missiles as Gromm was still atop the machine working on its Gatling guns. The tank rotated its upper thorax so fast to face the incoming threat that it sent Gromm flying several feet into the air, sending him smashing into the ground, breaking his neck upon impact.

The RAT engaged its anti-missile countermeasures, causing rapid-fire pulse cannons to rise from its back. The weapons locked onto the incoming rockets. But as soon as the cannons opened fire, they malfunctioned, just like its primary guns. Moments later, the swarm of missiles that weren't shot down collided with the RAT, causing it to explode in a bright mushroom cloud, spewing fire and debris in every direction.

The entire scene was sheer pandemonium. The handful of troops that survived Marcus' attack scattered in search of cover. However, the few that managed to make it inside the building were still locked in a life-and-death struggle with Skye and her crew. As they fought, Skye looked up from the gaping hole in the roof to see Marcus making strafing runs in *The Indicator*, forcing many of their would-be attackers to run in terror.

"It's Marcus," yelled Skye to the team, giving them a surge of hope, prompting them to fight even harder.

Tony dropped his granacite levels by more than half, sending him into an almost unstoppable frenzy. Three Lyrian attackers approached with spiked clubs and swords in hand. Before they could raise their weapons against him, Tony caught them with heavy blows, sending them crashing through the walls. He reached in the sheath strapped to his back and pulled out a machete the size of a short sword and went on the attack.

Shots fired and weapons clashed as Skye and her crew engaged the enemy with everything they had. Near the crumbling wall, Tommy pulled his heavy assault pistol from its holster, nailing attackers with perfectly aimed headshots. With the aid of his tactical shades and electronically augmented reflexes, Tommy acquired and engaged his targets with freakish speed and accuracy. The point-blank rounds from his handgun tore through the enemy's shields causing them to drop like ragdolls.

Across the room, Jason used a combination of his XP-90 pistol, cranked up to its maximum setting, and a metal rod he dislodged from the debris to fend off Sarrus' troops. And though he fought valiantly, the pain injector he had taken was starting to wear off, causing him to slow considerably. In short order, Jason found himself flat on his back, trying to keep his Lyrian opponent from plunging a sinister-looking dagger into his chest.

Completely out of ammo, Skye used her two curved blades to hack through her foes. And after severing the jugular of a human attacker, she looked over to see her brother in trouble. She hurled one of her blades at Jason's assailant, causing him to duck to avoid the knife. But before the Lyrian could reorient himself, Skye jumped on his back, causing them both to fall to the ground. But Skye maintained her rear chokehold with her left hand while hacking into her surprisingly resilient opponent with the blade in her right.

The Lyrian brute worked his way back to his feet, but Skye maintained her one-armed rear naked choke while continuing to impale the Lyrian with her knife. He thrashed around the room violently, slamming her into the crumbling walls, but Skye wrapped her muscular legs around his torso like a vice, locking in the hold even more. Despite the fatigue and pain, she continued to drive her bloodstained blade into her relentless adversary.

Exhausted and barely able to move, Jason struggled to overcome the intense pain raging throughout his body. The meds had fully worn off. He tried to help his sister but collapsed to the floor. While helpless on the ground, a Lyrian mercenary aimed her pistol at Jason to finish him off, but Tommy shot the weapon with his heavy assault pistol, obliterating her entire hand in the process. The force of the blast spun the Lyrian around to face Tommy, where he promptly drilled her between the eyes with a second shot from his prized weapon.

Skye continued hacking into her muscle-bound attacker like a woman possessed. It was a bloody affair, but she eventually wore him down until he fell to the ground, convulsing in a black pool of his own blood. As Skye rushed to her brother's side, Tony and Tommy finished mopping up the remaining Dragen Alliance troops.

After seeing Sarrus and his forces in full retreat, Marcus landed *The Indicator* in the open field just south of the building. He gathered his own weapons as well as the bag of supplies that included additional ammo, field rations, and canteens of water. He sprinted from the ship with the supplies, joining his old crew in what was left of the building. Once inside, he moved straight toward Skye. They rushed into each other's arms, setting aside all inhibitions, smothering each other with a kiss that took the rest of the crew by surprise.

Marcus dropped the supplies on the ground and the team descended on its contents like a pack of starving wolves. They practically inhaled the rations and water, having not had anything to eat or drink since they arrived on that infernal planet.

"Reload and get ready," said Marcus, prepping his own weapon. "It's not many of 'em left, but I'm sure they're regrouping for a counterattack." He turned to Tony. "Where'd they take Daren?"

Tony pointed to a large tower several miles down the road, beside which Marcus could see a large plume of black smoke.

"We managed to disable their transport," said Tony, pointing to his empty surface-to-air rocket launcher in the corner. "But I'm sure a backup is on the way."

As Tony spoke, Jason limped toward Marcus.

"What's the game plan?" said Jason.

But this time Skye wasn't having it.

"Ain't happenin' this time, boy," said Skye to her brother. "You can barely walk."

"She's right, Jay," said Marcus. "Think you can still fly?"

"Of course," said Jason, wondering why Marcus would ask such a stupid question.

"Good," said Marcus. "Make your way back to *The Indicator* and stand by to provide close air support." He took a long look at Skye, who looked almost as bad as her brother. "You go with him," he told Skye, "and get yourselves patched up while you're at it." He turned to Tony and Tommy. They both looked well enough considering the circumstances. "You two are with me."

Jason smiled. He relished the idea of fighting alongside Marcus once again and planned to make the most of every moment. As the siblings slowly made their way across the field to *The Indicator*, Tony and Tommy moved by Marcus' side, both eager for a little payback.

"Well, gentlemen," said Marcus to his small crew. "Shall we?"

Tommy and Tony nodded their heads in unison. And with that, Marcus and his two companions rushed down the muddy road toward Sarrus and his forces. It was time for round two.

CHAPTER FORTY-NINE

In the heart of the Citadel on Earth, the political situation was spinning out of control. President Jonathan Vance, flanked by Sarah Miller, General Thomas Kirkland, and other high-level advisors, huddled around the 90-inch view screen in the rear of his office. On the overlarge monitor, they watched live stealth reconnaissance footage of an even larger fleet of Lyrian warships than they had witnessed earlier. This time, the ships were conducting provocative military drills less than one parsec from ISL space, near the remote ISL Star Port, Stratus One.

"This is getting out of hand," said President Vance. "How can the GPC Council stand for this? Member races have never shown such aggression toward each other in the history of the Council."

"Well," said Sarah, shaking her head in dismay at the entire situation, "to let Lord Zek'Ren tell it, he *is* the Council."

"Those dirty reptilian-lookin' bastards," snarled General Kirkland. He turned to President Vance. "Admiral Athena Armstrong reports that the Onyx is on station at Stratus One, sir. Say the word, and we'll have the entire Seventh Fleet there in no time."

But President Vance remained silent, studying the situation playing out on the view screen before him.

"And I hate to be the bearer of bad news," said Sarah. "According to our high-level sources in the field, it appears that Ar'Gallious has made his move on the Benton Chamber as well. We don't know where exactly, but we do know it's happening."

"Get me something solid," said President Vance to Sarah. He turned to General Kirkland. "Anything from Captain Shepard?"

"Not yet, sir. I plan to check in with him after this meeting," replied the general.

"Well, don't let me hold you," said President Vance. The general nodded his head and started for the door, but President Vance momentarily stopped him. "General, do whatever it takes. I want Marcus La'Dek and his crew found."

President Vance turned to the rest of the staff and dismissed them, urging everyone to step up their efforts to bring the growing crisis to a swift conclusion; preferably one that didn't involve all-out war. As the last of his staff exited the room, President Vance pressed the communication button embedded in his desk.

"Mr. Paige," said President Vance to the executive assistant on duty. "Have Admiral Armstrong contact me directly."

President Vance turned his attention back to the view screen, deeply concerned about the way things were going down. "It might just be time to mobilize the fleet," said Jonathan to himself. He shut down the view screen and leaned back in his chair. *Sometimes I can't stand this job,* he thought. He reached for the bottle of liquor he kept stashed in a compartment beneath his desk and poured himself a drink, silently pondering his next move.

#

Inside La'Dek Transports on Stratus One, Donald excused himself to the empty office across from Samantha's to answer his Commanding Officer's call.

"What's going on, Captain?" demanded General Kirkland. "The president needs answers. And he wants them now."

"Nothing so far," said Donald, much to General Kirkland's irritation. "We've been questioning her for hours. But she knows more than she's telling us. I think she knows exactly where her brother is. So, how do you want to play it, sir?"

There was a long pause on the other end.

"Desperate times, Captain," said General Kirkland.

After hearing those three words, Donald knew exactly what the general had in mind.

"Understood, sir," replied Donald.

After terminating the communication, he turned toward the door and took a deep breath. *Alright, let's get this over with*, the captain thought. He straightened his uniform, opened the door, and marched back into Samantha La'Dek's office.

CHAPTER FIFTY

Locked in a pitched battle with Sarrus and his remaining forces on Xenon Prime, Marcus and his small team pressed the attack. Sarrus and his troops were not yielding ground, despite losing four of his seven remaining mercenaries. The rest of Sarrus' forces were dead, maimed, or had been completely routed from the battlefield. Marcus and his two-man crew took up positions behind a low stone wall. As they exchanged fire with Sarrus and his troops, Tommy looked over to Marcus.

"You guys take his team," yelled Tommy over the deafening sounds of battle. "Sarrus is mine!"

"No," yelled Marcus. "We have 'em right where we want 'em. We stick together."

But Tommy ignored Marcus and rushed toward Sarrus and his crew, attempting to outflank them. Furious and with half a mind to shoot Tommy himself, Marcus turned to Tony.

"Cover him," he yelled to the cyborg.

Tony laid down a wall of heavy suppression fire with is repeater, killing one of the three mercenaries, forcing Sarrus and his remaining two men to scramble for cover in different directions.

Seeing Sarrus isolated from his troops, Tommy approached from an angle that would have given him a clean shot. But in his haste, he forgot to reload his weapon. He squeezed the trigger, but there was no response. Not wanting to waste the opportunity, he rushed the Lyrian anyway, ready to take him out with his bare hands if necessary. But Sarrus picked up Tommy's approach, turning quickly to line up a shot.

Unable to complete his reload in time, Tommy hurled his weapon toward Sarrus, causing the shot to miss wide. Running as fast as he could,

Tommy took Sarrus down with a low tackle that nearly broke the Lyrian's leg.

Sarrus hit the ground with such force his weapon went flying from his hand, well beyond his reach. As the Lyrian lay on the ground, gripping his knee in pain, Tommy hopped on top of him in a full mount position, raining down punches with gloves lined with metal around the knuckles. After repeatedly whaling on his foe, Tommy wrapped his hands around Sarrus' throat, trying his best to end the Lyrian's life for that stunt he and his brother pulled back on Protos IV.

"Thought you could get one over on me?" yelled Tommy, trying with all his might to crush the Lyrian's windpipe.

Before Tommy could finish him off, Sarrus turned the tide with his superior strength, managing to roll Tommy off him. With his enemy flat on his back, Sarrus hopped on top, where he launched a vicious assault of his own, nearly knocking Tommy cold. As Sarrus reared his arm back to punch again, he deployed the talons of Drakos, causing the razor-sharp blades to extend from his gauntlets, like the claws of a ferocious predator.

Concerned that he might hit Tommy by mistake, Marcus rushed as fast as he could to help, tackling Sarrus to the ground. He quickly scrambled to his feet, pulling his XP-90 from its holster, ready to blast the Lyrian's head from his shoulders. But he was too close. And before Marcus could squeeze the trigger, Sarrus swiped at the handgun with his talons, shredding it into pieces, nearly taking Marcus' hand off in the process. Dropping what was left of his weapon to the ground, Marcus went on the attack.

Despite his skilled use of the Talons of Drakos, Sarrus' attacks were off balance and mostly ineffective due to his injured leg. Marcus seized the advantage by making himself an elusive target, using Sarrus' momentum against him, repeatedly counter-punching and slamming the large Lyrian to the ground.

Moments later, Tommy finally recovered, staggering back to his feet. To his left he saw that Tony had the two remaining Lyrian mercs pinned. To his right, he saw the brawl between Marcus and Sarrus. With rage in his eyes, Tommy rushed toward Sarrus, delivering a kick to the chest that knocked the Lyrian off his feet, sending him crashing to the ground.

Marcus and Tommy rushed toward their fallen opponent and proceeded to beat him to within an inch of his life, repeatedly kicking, punching, and stomping the Lyrian into the mud-covered grounds of Xenon Prime. After finishing off the final two Lyrian mercs, Tony joined Marcus and Tommy. When he arrived, he found Sarrus twitching on the ground, barely able to move.

"Looks like I missed out," said Tony.

"Think this is over?" grunted Sarrus. "You have no idea what's—"

Before he could finish, Tommy grabbed the XP-90 from Tony and emptied an entire clip into the torso of Sarrus, silencing his annoying voice, permanently. Tommy ejected the spent magazine and tossed the weapon back to Tony.

"Thanks, Tony," said Tommy.

"Don't mention it."

"Let's move," said Marcus to his crew.

They sprinted toward the ominous tower in the distance, where Goran was no doubt holding Daren.

#

As they awaited further instructions from Marcus, Skye used the short downtime in *The Indicator* to tend to the worst of Jason's injuries, after which she patched up a few of her own. She was relieved to have access to decent medical supplies for a change.

"Congratulations, Jay," said Skye. "You'll live to fight another day."

Jason was overwhelmed at how his sister saved his life. But he was even more shocked when she revealed that Tommy also saved his life, twice in the same day. As highly intelligent and well-spoken as Jason was, he was horrible at expressing gratitude. And while he would have found it especially difficult to thank the likes of Tommy, he was relieved that he could at least start with his sister.

"Umm," said Jason, not knowing how or where to even start. "What you did for me back there..."

But Skye interrupted his speech with a heartfelt hug.

"Save it, Jason," said Skye to her younger brother. "I can't let anything happen to you on my watch. Mom would kill me."

The two laughed for a moment. They made a pact with each other that once this was all over, they would reconnect with their mother, whom Jason hadn't seen since his father took him away from her as a child to live on Earth, and whom Skye hadn't seen in person in years.

As they continue to prep the ship, the call they were waiting for finally came through.

"Jason! We're headed to the tower," yelled Marcus over the comm unit. "Get that thing airborne and clear us a path."

Jason acknowledged. His leader's words were like music to his ears. With a renewed surge of energy, he fired up *The Indicator*'s engines.

"You heard the man," said Jason to his sister. "Let's hit it."

Skye smiled, seeing a spark in her brother's eyes once again. She strapped herself into the copilot's chair and readied the ship's weapons. Within minutes, *The Indicator* launched vertically from the ground, adjusting its heading toward the tower. In seconds, the ship rocketed toward its target with a sudden burst of speed.

#

As Marcus and his team approach the tower, they could hear the rumble of *The Indicator*'s engines as it raced over their heads. Moments later, they could see *The Indicator* raining fire down upon the forces guarding the entrance. The enemy had no significant anti-air capabilities, so they never really stood a chance.

"You're all clear," said Jason over the radio.

By the time Marcus and his team arrived at the lofty structure, all they saw around them was death and destruction. Everywhere they looked, there was nothing but twisted, smoldering heaps of what used to be anti-personnel cannons, and trails of dead or severely injured guards. As Marcus and his crew stormed the building, Jason put *The Indicator* into a holding pattern, causing it to hover near the roof of the tower.

Less than twenty minutes later, emergency alerts blared throughout the bridge of *The Indicator*.

"We've got incoming," yelled Skye. "Picking up two ships, a Dragen transport and a fighter."

Jason initiated evasive maneuvers.

"Looks like their ride is here," said Jason as he activated the ship's afterburners, narrowly dodging the incoming cannon fire from the Dragen Alliance fighter.

As the fighter pursued *The Indicator*, the transport landed on the roof of the tower, after which a four-Lyrian fire team of elite Dragen Alliance soldiers exited the ship. They wasted no time sprinting toward the stairwell entrance on the rooftop. The silver fighter was a nimble ship with sharp, angular features. While not heavily armed, the ship was fast, and packed enough of a punch to give *The Indicator* a run for its money, especially in *The Indicator*'s current condition.

While impressed with what Marcus was able to do with the backup weapon system, Jason knew that without *The Indicator*'s primary armament, the backup system would be minimally effective against such a fast mover. So, he spent most of the time dodging the relentless barrage coming from the fighter. And the few times he managed to fire on his target, he missed spectacularly.

The stress on *The Indicator*'s propulsion system, caused by Marcus' omega-jump, was starting to rear its ugly head. With every twist and turn Jason made, he fought just to keep *The Indicator*'s engines operational. Moments later, several rounds from the fighter pierced *The Indicator*'s shields, taking out its left main engine.

"We're goin' down," yelled Jason. He attempted to level out *The Indicator*'s rapid descent as they thundered toward the ground behind a distant hill. "Brace for impact!"

He used *The Indicator*'s remaining engine, as well as its maneuvering thrusters, to slow the ship enough to keep it from exploding on impact. But they were out of the fight. While it was a rough landing, Skye and Jason survived the crash. And to their surprise, *The Indicator* held together better than expected. But they wouldn't be leaving the surface of Xenon Prime until they made significant repairs to that left main engine.

It was clear from the moment they crashed that the fighter's primary mission was to drive them away from the tower, giving the transport enough time to pick up Goran and Daren. And after successful completion of that mission, the fighter broke off its assault and turned back to orbit the tower.

#

Inside the tower, Marcus and his team had cleared out the remaining guards and managed to chase Goran to the top floor. He stood near the stairwell door using Daren, whose hands were bound behind him, as a human shield.

"Let him go, and hand over the chamber, Goran," said Tommy, motioning toward the large black duffel bag, containing the portable Mk17, strapped to Goran's back. "And I might spare your life. Unlike I did your brother back there!"

"Sarrus..." said Goran, clearly distraught. The reality that his brother had died suddenly hit him like a hover train, sending him into a fit of rage.

"You'll burn by the Breath of Drakos for this, Lance!" roared Goran to Tommy.

"Yeah, whatever," said Tommy. "You got five seconds."

"It's over, Goran," said Marcus. "There's nowhere to run this time."

In that moment, Marcus and Daren briefly locked eyes, signaling for Daren to make a move.

Daren drove the back of his head into Goran's face, creating enough space for Marcus to get a shot off, hitting Goran square in the shoulder, though he was aiming for his head. Before the rest of the team could open fire on Goran, the stairwell door flung open, followed by a storm of energy bolts coming from the rifles of the four Dragen Alliance soldiers who had landed on the roof.

As soon as they fully entered the room, the soldiers tossed down what looked like a long chain of metal bars. From the metallic rods, a phalanx of glowing energy shields emanated, forming a five-foot-tall barrier of raw power between Marcus' team and the Lyrian soldiers, including Goran and Daren.

Daren scrambled to his feet to make a run for it, but as soon as he stood, he was shot in the back with a shock round by one of the Lyrian guards, totally incapacitating him. He fell to the ground in a fit of violent muscle spasms, then blacked out.

Marcus and his crew ran for cover and returned fire. But their rounds were unable to pierce the pulsating green barriers. Even the overcharged rounds from Tony's heavy repeater were easily absorbed into the chain of energy shields like water in a sponge. Tony ceased fire and low-crawled behind cover toward Marcus.

"There's nothin' we can do. I've seen those shields before; we'll run out of ammo before we pierce those things," said Tony, irritated that he was even suggesting retreat. But he'd seen enough battle in his time to know when it was time to back off.

Marcus was overwhelmed with emotion, being so close to making everything right. He was desperate to rescue Daren and recover the Benton Chamber, hoping to avoid a war that would surely affect his family, not to mention the lives of trillions across the known galaxy.

"Cover me," said Marcus. Without thinking, he attempted to run straight into the line of fire to go for Daren. But Tony grabbed Marcus by the arm, preventing him from rushing.

"It's suicide, man," said Tony. "We need to fall back."

Marcus snapped out of it, realizing that Tony was right. He ordered the retreat, vowing to Daren as they fell back that he would come for him.

During the firefight, one of the guards grabbed Daren, hoisting him on his shoulders, and retreated through the stairwell toward the rooftop. Still clutching his bleeding shoulder, Goran secured the duffel bag containing the Benton Chamber, and followed the guard toward the transports waiting on the roof. When the soldiers saw Marcus and his crew backing off, they ceased fire and sprinted up the stairwell behind Goran.

On the rooftop, the Lyrian soldier carrying Daren placed him in the transport, this time binding his legs in shackles. Once Goran entered the ship, the awaiting medic went to work on his shoulder. After the remaining three guards entered the transport, the ship lifted from the rooftop. In seconds, the transport blasted off toward space, followed by the fighter that had been circling the tower.

From the ground, Marcus and his crew looked up to see the ships ascending into the clouds. Marcus shook his head in disappointment, then contacted Jason on comms.

"They got away with Daren," said Marcus.

There was a long pause on the other end, then Jason responded.

"Roger that," said Jason, clearly disappointed.

"We're ready to head back," said Marcus. "Where are you?"

"Grounded for now," said Jason while working feverishly to repair the damage from the skirmish with the Lyrian fighter. "They took out one of the engines. But I could use a hand with these repairs. Transmitting our location to you now."

"This day just can't get any better," said Marcus to himself. After receiving *The Indicator*'s position, Marcus replied to Jason. "Location received. We're on our way."

CHAPTER FIFTY-ONE

Hours later, Marcus and most of the crew were hard at work making repairs to *The Indicator*'s engine. It was a somber mood as they all worked in silence. While they hadn't discussed the events that had occurred on the rooftop since their return to the ship, Tony could see that it was weighing heavily on Marcus. He stopped what he was doing and approached his friend and leader, putting his cybernetic hand on Marcus' shoulder.

"You did all you could out there, brother," said Tony.

But Marcus didn't respond.

"We'll get him back," Tony continued.

Marcus appreciated Tony's words, but before he could respond, Skye rushed into the engine room.

"Marcus," yelled Skye. "Come to the bridge quick. There's an emergency audio-trans from Stratus One. It's somethin' about your sister."

Marcus, Tony, Jason and Skye sprinted toward the bridge, leaving Tommy behind to continue the repairs alongside the maintenance droid. Once on the bridge, Marcus answered the call. It was Kayla. She was calling from her personal communicator from La'Dek Transports. She was hysterical.

"I don't know what to do," said his cousin, sounding as if she was in tears.

"Calm down, Kayla," said Marcus. "What happened?"

#

As Kayla spoke to Marcus inside the empty back office, armed soldiers ransacked the front-desk area, accessing every computer terminal in the room. As they wrecked the office, Elizabeth was crying in the corner, calling out for her mother. But her voice fell on deaf ears as the calloused soldiers continued tearing the place apart, refusing to let her near her mother. The entire scene crushed Kayla's heart. Barely able to speak, she turned back to her comm unit.

"Kayla!" yelled Marcus on the other end. "What's goin' on?"

"They arrested Sam!" she screamed. "She told me to contact you on this channel if something happened..."

Before she could finish, one of the soldiers stormed into the room. They tussled as the guard tried to confiscate her communicator, but in an act of defiance, Kayla smashed her comm unit into pieces, rendering the device useless.

"Who were you talking to?" yelled the soldier.

"You get the hell out of our office," said Kayla, raising her voice in kind. "You have no right—"

The guard shoved her to the ground and dragged her from the back room into the front office. From across the room, Samantha stood in tears next to Donald, her wrists bound in shackles. She tried desperately to keep a bad situation from becoming worse. Having worked with soldiers such as these throughout her military career, Samantha knew the lengths to which Donald and his team would go in search of answers. And with no desire to see any of her family hurt, she raised her voice.

"Kayla. Elizabeth. Stop," she yelled to her family. "Just let these soldiers do their jobs. We have nothing to hide," she continued, glaring at Donald.

Elizabeth calmed herself and approached the Captain, but Lieutenant Alex Chavez intercepted her.

"Back away, kid," said Alex. "Do what your mother said."

Elizabeth wasn't having it and challenged the officer.

"Under whose authority are you even here?" Elizabeth yelled. "Section three, paragraph seventeen of ISL criminal law states that we have rights. You have to disclose the full list of charges brought against the accused. It also says—"

"I said *step back*," growled Alex, incensed by the audacity of Elizabeth, and shocked by her level of knowledge of the law. "We ain't the Interstellar Police... This is a military matter, so for your own safety—"

"Is that so, *Lieutenant?*" said Elizabeth, mocking the officer. "Because even the ISG Code of Military Justice articles that govern *your* interactions with civilians afford us similar protections as—"

Fed up with the girl's insolence, Alex stood face to face with Elizabeth. As he did so, Samantha rushed toward Alex despite her shackled hands, shoving the soldier away from her daughter.

"You have a problem with *my* child, you talk to me, Lieutenant," shouted Samantha, sounding as if *she* was *his* commanding officer.

Alex moved toward the prisoner, ready to put her in her place. But Donald intervened.

"Stand down, Lieutenant," said Donald in a calm voice. "Have your team wrap this up so we can go. We're headed to the ship."

"Yes, sir," said Alex forcefully. He wasn't accustomed to his actions being questioned by a civilian, let alone by a know-it-all seventeen-year-old. However, he couldn't allow his troops to see him lose his cool. So, he took a deep breath and turned to two of the five soldiers searching the office.

"Corporal Tannahill, Specialist Carter," he said, summoning the two soldiers. "We're done in here. I need you two to search the terminal in Ms. La'Dek's office, then fall back to the ship," said Alex. He rallied the rest of the troops and followed Donald, including their shackled prisoner, out the door toward the Stratus One docking bay, leaving a devastated Elizabeth and Kayla behind in his wake.

Corporal Tannahill and Specialist Carter wrapped up their search in the front office then started toward the back, but Elizabeth shoved her way past them and ran to her mother's office, planting herself in front of Samantha's computer. Tannahill and Carter entered the office, shutting the door behind them. They chuckled at Elizabeth's pathetic attempt to block their access.

"Look, Elizabeth, we don't have time for this," said Tannahill, trying to sound as professional as she could. "The sooner you get out of the way, the sooner we can leave."

"Yeah," said Carter, trying to suppress his laughter. "Just stand aside so we can get this over with."

"No," yelled Elizabeth, shaking in anger.

The mood in the room turned serious. Tannahill stepped forward, deciding to talk to Elizabeth woman to woman.

"Look, kid," said Tannahill as she slowly approached, "I understand how you feel—"

"You have no idea how I feel right now," said Elizabeth, her voice starting to shake.

Irritated and ready to go home, Tannahill moved closer.

"Step away from the computer, or we'll have to move you," said Tannahill, causing her partner to become uncomfortable with the unexpected standoff.

"I'd like to see you try," said Elizabeth, almost sounding like another person. "Get out of this office."

"The hard way then," said Tannahill, much to her partner's dismay.

The two soldiers reluctantly moved forward, but as they approached, they froze in their tracks with fear. Without warning, Elizabeth's eyes darkened, emitting a purplish-black glow. A shadowy aura of mist surrounded the girl, darkening her visage.

"I said get out!" screamed Elizabeth.

As she spoke, the temperature in the room dropped several degrees. Moments later, a dark energy radiated from her body and slammed into the soldiers. The force of the blast sent them flying across the office, temporarily pinning them against the wall. The pressure of the unseen power was so great, the soldiers could barely breathe.

And just as quickly as the event happened, Elizabeth's countenance returned to normal, and the mysterious energy dissipated, causing the soldiers to fall to the floor.

Drained by the experience, Elizabeth also collapsed to the ground, unconscious.

"What the hell was that?" said Carter, struggling to stand to his feet, his face still pale with fear.

"I don't know," said Tannahill, shaking uncontrollably as she watched Elizabeth lying on the ground.

"We'll tell 'em the computer was clean. Let's just get outta here," said Carter.

Tannahill agreed, and the two petrified soldiers sprinted from the office as fast as they could, vowing never to tell anyone of the events that had just transpired, fearing they would be labeled as crazy. It would mean the end of their budding military careers. But one thing was certain. The events in that room would be indelibly etched in their memories, haunting their dreams for the rest of their days.

At the Stratus One docking bay, Donald and Alex had just marched Samantha into their awaiting ship. Minutes later, Tannahill and Carter caught up with the group, still shaken by their encounter with Elizabeth. However, when the two soldiers checked in with Alex, they falsely reported their task as complete, wanting only to put as much distance between them and that star port as possible.

Inside the transport, Samantha looked out the starboard-side viewport with tears in her eyes, wondering if she would ever see her family again, especially her daughter, Elizabeth. After a few moments, she wiped the tears from her eyes and sat up straight, refusing to give Donald and Alex the pleasure of seeing her distraught.

Inside La'Dek Transports, Kayla stayed by Elizabeth's side in her unusually weakened state, trying her best to be strong for her cousin.

#

On the bridge of *The Indicator*, silence fell over the team the moment the communication with Kayla had abruptly terminated. Skye took a seat next to Marcus in the co-pilot's chair. She turned to Jason.

"Go help Tommy finish the repairs so we can get off this rock," said Skye.

Jason nodded then rushed to the engine room. Skye turned to face Marcus.

"We'll get this bird flyin' again," she said.

Moments later, Tony stepped forward.

"We're gonna make this right," said Tony. "All of it."

"We're all in, Marcus," said Skye. "'Til the end."

Marcus believed them. And in the face of such a crisis, he could think of no other group he'd rather be with to weather the storm. The team *was* part of his family, and in that moment, he knew that they would go to the ends of the galaxy for him and for those he loved back home. Marcus sat back in his chair for a moment, then arose from his seat. He turned toward his companions.

"Then I guess we have our work cut out for us," said Marcus. He started to move toward the engine room to assist in the repairs, but the cyborg stopped him.

"You stay here," said Tony. "We'll handle it."

Skye nodded in agreement and moved face to face with Marcus. She put one hand on the side of his face, then kissed his cheek.

"We'll let you know when it's done," said Skye.

Marcus nodded and sat down in the captain's chair. After his companions left the bridge for the engine room, he leaned back in his seat, staring into the blank view screen. All he could think about was his sister and the rest of the family on Stratus One. But his niece, Elizabeth, weighed the heaviest on his mind. He wanted to be there for her in the wake of her mother's arrest. There was so much he needed to say.

Since he couldn't be there, he decided to put his thoughts in writing. In truth, Marcus rather enjoyed writing, though he hadn't done it much since his release from prison. Though he dropped out of school to run off with Daren and his father to learn the mercenary trade, Jax always stressed academics as much as weapons training, wanting to ensure that Daren and Marcus could operate in any social circle.

Instead of dictating his words to the computer, he opted to type his letter instead. He moved from the captain's chair and sat at the workstation near the rear of the bridge. With the wave of his hand over the communication console, a holographic keyboard appeared with a cursor that seemed to float in midair. After taking a moment to gather his thoughts, he began to type.

About midway through his letter, he caught a glimpse of the newsfeed running silently on the monitor to his left. On the screen was leaked footage of a massive buildup of Lyrian warships just one sector from Stratus One, followed by footage of an armada of Interstellar League warships in another sector. At the forefront of the Interstellar Fleet was its flagship and crown jewel, *The I.S.F. Onyx*, commanded by the legendary Admiral, Athena Armstrong. Seconds later, a caption flashed across the screen, *EMPIRES ON THE BRINK OF WAR?*

Marcus unmuted the volume to hear the voice of Bobby Wiseman, the well-known reporter for the Galactic News Agency.

"With no progress in the search for Dr. Jakob Benton and his groundbreaking prototype dubbed the Benton Chamber, tensions between the ISL government and the Lyrian empire are reaching a boiling point." As the reporter spoke, the footage switched from the fleet buildup to an image of the Benton Chamber. "According to our sources embedded in the Empire, the Lyrians are convinced that Dr. Benton's invention isn't missing at all, but that ISL President Jonathan Vance is purposefully withholding the device for nefarious reasons, an act which the Lyrian Emperor, Zek'Ren, says 'will have dire consequences.'"

The image on the screen faded away, focusing back to the silver-haired reporter, a concerned look on his face. "For those who don't live on terraformed worlds and may not understand why this is such a big deal, allow me to explain: When Krillium ore supplies run dry within twenty Earth years, every terraformed planet in the galaxy will experience total atmospheric collapse, resulting from the loss of power of their Terraforming Modules. And without the Benton Chamber to harness the immense energy output of the replacement power source, a newly discovered crystal called Ladium, there may be no way to prevent this looming catastrophe."

Marcus had heard enough. He shut down the monitor, staring for a moment into the blank screen. His heart sank, knowing that their actions directly contributed to the growing crisis. *This thing is way bigger than my family*, Marcus thought, realizing that their actions may have galaxy-wide repercussions. As he sat in silence, a fire burned within. He was determined more than ever to fix the mess he and his crew had caused. He took another deep breath, turned back to his terminal, and continued to type.

Upon finishing the letter to his niece, Marcus looked it over, just to make sure it was perfect. As he did so, Skye contacted him from the engine room, notifying him the repairs were complete. He could hear the rest of the crew celebrating in the background. Marcus leaned back in his chair, letting out a sigh of relief.

Moments later, the crew, including Tommy, entered the bridge in high spirits, relieved that they could finally leave Xenon Prime. Even Tommy and Jason agreed to bury the hatchet for the time being. They shook hands as Jason thanked Tommy for saving his life earlier that day.

Marcus looked on with approval.

After taking a seat in the captain's chair, Jason fired up *The Indicator*'s engines, and launched the ship from the planet's surface toward space. Once underway, Marcus turned his attention back to the letter. Satisfied with his words, he hit the submit button, sending the heavily encrypted message to Elizabeth's personal communicator. With everything going on, he figured it might be awhile before she actually got around to reading the letter. But he had to get it off his chest anyway. Marcus leaned back in his chair and took a deep breath.

"Alright, folks," said Marcus, swiveling around in his chair to face his crew, "let's do this."

And with that, Jason engaged *The Indicator*'s hyper-jump engines, causing *The Indicator* to jump into hyperspace in a radiant burst of light.

EPILOGUE

Mentally and physically drained, Elizabeth lay sprawled across her bed, staring blankly at the ceiling of her cramped bedroom on the Stratus One Star Port. She had no more tears to cry and was tired of her cousin's attempts at cheering her up. She just wanted to be alone. But all she could do was think about her mother and uncle. She felt helpless. And for all the advanced knowledge floating around in her head, she had no idea what to do. Just then, the soft chime of her personal communicator interrupted the silence, indicating there was an incoming message.

She checked the device to find a strangely encrypted transmission attached, with seemingly no way to open the file. She started to delete it so she could get back to sulking, but the subject line gave her a moment of pause. It said, *bet you can't crack this one.* Elizabeth smiled, convinced that the anonymous transmission was definitely from her uncle.

Marcus was fascinated with Elizabeth's freakish intelligence and was always trying to test the limits of her abilities. So, they had a game they would sometimes play. Marcus would send her a random secure message, daring his niece to break into the file. Of course, it was mere child's play for Elizabeth, but she enjoyed playing along just the same. She would quickly scythe the file, just to see whatever silly message or image her uncle decided to send.

But the encryption level on the transmission she just received was beyond military grade, causing her to become even more intrigued. With everything that had just happened, she was grateful for the mental diversion. She transferred the file from her communicator to her custom-built system on the messy desk adjacent to her bed. Using the system's holographic keyboard, she wrote a decryption program from scratch, as her previously written programs would have no chance of bypassing the security measures on the file. It took a little more thought on her part than usual, but she ultimately cracked the code. She was right. It was a letter from Marcus, dated *Earth year 3120*. With fresh tears welling in her eyes, she read the following words...

Dearest Elizabeth,

I know exactly how you feel. Believe me, I'm feeling the same way. And while I can't go into all the details right now, you deserve to know the truth. And the truth is, this is all my fault. The word "sorry" can't begin to erase the pain I've caused. What can I say? I let my past failures impact my present. And I made some really bad decisions along the way. And now those decisions have come back to affect all of you. And for that, I am truly sorry.

An old lady once told me years ago that lost time is never found again. I guess it was her way of telling me that you can't go back. We can't go back to recover the years we sometimes waste in life. And I'll be the first to say that I've wasted many years, which kept me away from all of you. But thankfully, that's not the end of the story. The lady went on to tell me that a failed past doesn't have to be a death sentence for your future. But it's what we do now that matters.

In truth, we all carry precious cargo through life. And that cargo is everything from the choices we make to the relationships we forge. But it is the love and compassion we have for one another that are the ties that bind us together. In the end, it all helps to define who we are, and says a lot about where we're headed.

Your mother once told me that you can't choose family, whether we share the same blood or not. And when adversity comes, we have to roll with the punches. Because hey, that's what we do.

I know this is the last thing you want to hear, but before I come home, there's something I have to do. Think of it as my first step in fixing this huge mess I've made. But don't worry.

Kayla knows the business like the back of her hand and will take good care of you while me and your mother are away. So, be sure to help your cousin out, you hear? I love you and I promise to check in as often as I can. Take care, and I'll see you all soon.

With Love,
Uncle Marcus.

P.S.

Tell Kayla I love her too, and that I said thank you. And I promise you both, I'll make all of this right.

After reading the letter, Elizabeth closed her eyes and exhaled, finding some small measure of peace. Though there were so many unanswered questions, she took solace in knowing that her uncle was still alive and would stop at nothing to fix their broken family. And while she had no idea what tomorrow would bring, she said a prayer for her mother and uncle. She finally closed her eyes and slept, confident that God would protect her family and would one day bring them all back together.

# # #

www.ingramcontent.com/pod-product-compliance
Lightning Source LLC
Chambersburg PA
CBHW031842200326
41597CB00012B/232